Advanced Simulation and Modelling for Urban Groundwater Management – UGROW

Urban Water Series – UNESCO-IHP

ISSN 1749-0790

Series Editors:

Čedo Maksimović

Department of Civil and Environmental Engineering
Imperial College
London, United Kingdom

Alberto Tejada-Guibert

International Hydrological Programme (IHP)
United Nations Educational, Scientific and Cultural Organization (UNESCO)
Paris, France

Sarantuyaa Zandaryaa

International Hydrological Programme (IHP)
United Nations Educational, Scientific and Cultural Organization (UNESCO)
Paris, France

Advanced Simulation and Modelling for Urban Groundwater Management – UGROW

Dubravka Pokrajac

School of Engineering, University of Aberdeen, UK

and

Ken Howard

Department of Physical and Environmental Sciences, University of Toronto at Scarborough, Toronto, Canada

UNESCO
Publishing

United Nations
Educational, Scientific and
Cultural Organization

CRC Press
Taylor & Francis Group
Boca Raton London New York

CRC Press is an imprint of the
Taylor & Francis Group, an **informa** business

A BALKEMA BOOK

CRC Press
Taylor & Francis Group
6000 Broken Sound Parkway NW, Suite 300
Boca Raton, FL 33487-2742

First issued in paperback 2019

© UNESCO, 2010
CRC Press is an imprint of Taylor & Francis Group, an Informa business

Typeset by MPS Limited (A Macmillan Company), Chennai, India

No claim to original U.S. Government works

ISBN UNESCO, paperback: 978-92-3-104173-0
ISBN-13: 978-0-415-45354-7 (hbk)
ISBN-13: 978-0-367-38335-0 (pbk)

Urban Water Series: ISSN 1749-0790

Volume 7

The designations employed and the presentation of material throughout this publication do not imply the expression of any opinion whatsoever on the part of UNESCO or Taylor & Francis concerning the legal status of any country, territory, city or area or of its authorities, or the delimitation of its frontiers or boundaries.
The authors are responsible for the choice and the presentation of the facts contained in this book and for the opinions expressed therein, which are not necessarily those of UNESCO nor those of Taylor & Francis and do not commit the Organization.

British Library Cataloguing in Publication Data
A catalogue record for this book is available from the British Library

Library of Congress Cataloging-in-Publication Data

Pokrajac, Dubravka.
  Urban groundwater systems modelling / Dubravka Pokrajac, Ken Howard.
     p. cm. – (Urban water series–UNESCO-IHP, ISSN 1749-0790 ; v. 7)
  Includes bibliographical references.
  ISBN 978-0-415-45354-7 (hardback : alk. paper) – ISBN 978-0-415-45355-4 (pbk. : alk. paper) – ISBN 978-0-203-94705-0 (ebook)
  1. Groundwater flow–Simulation methods. 2. Urban runoff–Simulation methods. I. Howard, Ken W. F. II. Title.
  GB1197.7.P65 2010
  551.4901'1--dc22
                          2010031758

**Visit the Taylor & Francis Web site at
http://www.taylorandfrancis.com**

**and the CRC Press Web site at
http://www.crcpress.com**

Foreword

Groundwater represents the largest reserve of available freshwater resources, providing a reliable source of water for human use. A large part of the world's urban population today depends on groundwater for its drinking water. In general, groundwater is of a good quality – superior to that of surface water – and has modest treatment requirements. Because groundwater resources can be developed easily and with relatively low costs, they are used extensively for domestic, industrial and agricultural water supply throughout the world. The importance of groundwater in urban areas is recognized not only as a valuable source of good quality water, but also for its role in the urban water cycle and in mitigating the impact of drought on urban water systems. Groundwater resources in and around most urban areas are under threat as a result of over-exploitation and degradation due to pollution.

This book aims to contribute to effective management of groundwater resources in urban areas, using simulation and modelling tools to better understand groundwater interactions with urban water systems. It presents a simulation and modelling tool, *UGROW* – one of the most advanced urban groundwater modelling systems developed to date – which integrates all urban water system components. The *UGROW* model was developed to support decision-making in urban water management and was tested using different case studies on urban groundwater.

In an effort to advance the scientific knowledge for better management of groundwater resources in urban areas, the Sixth Phase of UNESCO's International Hydrological Programme (IHP-VI, 2002–2007) implemented a project with a specific focus on urban groundwater modelling. The project was implemented with the support of a working group, composed of experts in the areas of urban water, hydrogeology and modelling, whose valuable contributions and collaborative effort through a series of workshops resulted in the publication of this book. The contribution of Dubravka Pokrajac, University of Aberdeen (UK), and Ken W.F. Howard, University of Toronto (Canada) was indispensable as the editors and main contributors of the book and is amply acknowledged.

Published in the UNESCO-IHP Urban Water Series, this book was prepared under the responsibility and coordination of J. Alberto Tejada-Guibert, Deputy-Secretary of the IHP and officer responsible for the IHP's Urban Water Management Programme, and Sarantuyaa Zandaryaa, Programme Specialist in urban water management and water quality at UNESCO-IHP, whose efforts as co-editors-in-chief of the series were central in the publication of the book. The role of Čedo Maksimović, Imperial College (UK), as a co-editor-in-chief of the series is equally acknowledged with appreciation.

UNESCO extends its gratitude to all contributors for their outstanding effort, and is confident that the concepts and novel ideas presented in this book will be of value to urban water management practitioners, policy-makers and educators alike throughout the world. The publication of this book is a major contribution to enhancing the scientific knowledge for sustainable urban water management. It will contribute to a better understanding of groundwater resources in urban areas.

International Hydrological Programme (IHP)
United Nations Educational, Scientific and Cultural Organization (UNESCO)

Contents

List of Figures

List of Tables

List of Acronyms

3D	three-dimensional
3DNet-UGROW	a UGROW integrated hydro-informatics tool
AISUWRS	Assessing and Improving Sustainability of Urban Water Resources and Systems
ARCINFO®	a GIS software
ASCII	American Standard Code for Information Interchange
ASR	Aquifer Storage and Recovery
CCTV	Closed Circuit Television
CNTB	Central Northern Tampa Bay area
CSIRO	Commonwealth Scientific and Industrial Research Organisation
DAG	Directed Acyclic Graph
DELINEATE	a UGROW component used for determining catchment areas
DHI	Danish Hydraulic Institute
DHI-WASY GmbH	the German branch of the DHI group
DSS	Decision Support System
DTM	Digital Terrain Model
ESI	Environmental Systems International
EU	the European Union
FEFLOW®	a finite-element groundwater model
GEOLOGY	a UGROW component used for handling geological layers
GEOSGEN	UGROW algorithm for generation of 3D solids
GIS	Geographical Information System
GO	Graphical Objects
GRID	a GIS reference for handling data
GROW	UGROW simulation model of groundwater flow
GTA	Greater Toronto Area
GUI	graphical user interface
HSPF	Hydrologic Simulation Program – Fortran
ICU	Intermediate Confining Unit
IHP	UNESCO's International Hydrological Programme
ISGW	Integrated Surface water Ground Water software
IWRM	Integrated Water Resources Management
MESHGEN	UGROW algorithm for mesh generation
MKL	Middle Gravel Layer

MODFLOW	a finite-difference groundwater simulation model
MODFLOW-SURFACT	a model which incorporates flow in the unsaturated zone, delayed yield, and vertical flow components
MODPATH	a 3D particle-tracking model
MS-Access	a database software
MT3D	a 3D contaminant transport model
NAPL	non-aqueous phase liquid
NEIMO	Network Exfiltration and Infiltration Model
OKL	Upper Gravel Layer
OROP	Optimized Regional Operations Plan
PEST	parameter estimation and automatic calibration software code
POSI	a purpose-designed unsaturated zone model
PSLG	Planar Straight Line Graph
QA	Older Quaternary
SAS	Surficial Aquifer System
SDI	SDI Environmental Services, Inc., a hydrological consulting services firm
SEAWAT	a 3D, variable-density, transient groundwater flow model
SEWNET	part of WATER which handles urban drainage
SI	International System of units
SLeakI	a purpose-designed unsaturated zone model
STREAMNET	part of WATER which handles urban streams
TIN	Triangular Irregular Network
UFAS	Upper Floridan Aquifer System
UFIND	UGROW algorithm for assigning sources of recharge to individual mesh elements
UGROW	Urban GROroundWater modelling system
UL_FLOW	a purpose-designed unsaturated zone model
UNSAT	UGROW simulation model of flow in unsaturated zone
USDA-SCS	United States Department of Agriculture-Soil Conservation Service
USGS	United States Geological Survey
UTM	Universal Transverse Mercator
UVQ	Urban Volume and Quality model
UWP	Urban Water Programme
WATER	a UGROW component used for operating all water systems
WATNET	part of WATER which handles water supply systems
W-E	West–East

List of Symbols

a	coefficient
as	subscript indicating the areal source number s
a_i	coefficient of the polynomial number I, unknown time-dependent coefficients
a_i^{new}	new value of H at the computational node
a_i^{old}	old value of H at the computational node
a_s	geometry of the area defining the position of the areal source number s
b	coefficient
d	depth to water table
dt	infinitesimal time interval
dx	infinitesimal element of x
dy	infinitesimal element of y
dW	infinitesimal element of W
f	function defined along a side of a finite element
f_j^q	terms on the right-hand side of the j-th equation
f_j^B	terms on the right-hand side of the j-th equation
g	acceleration due to gravity
h	capillary pressure head
h_0	reference pressure head
h_1	capillary potential at the soil surface
h_j^{k+1}	pressure head at a node j and at the time level t^{k+1}
h_p	prescribed depth of water layer on the soil surface
i	subscript denoting the direction of a Cartesian coordinate
$\vec{i}$	basic unit vector of the x coordinate
j	subscript denoting the direction of a Cartesian coordinate
$\vec{j}$	basic unit vector of the y coordinate
k	resistance coefficients for infiltration/exfiltration, hydraulic conductivity of the unsaturated soil
k_0	reference leakage parameter
$\vec{k}$	basic unit vector of the z coordinate
k_a	resistance coefficients for infiltration/exfiltration
k_b	resistance coefficients for infiltration/exfiltration

k_c	resistance coefficients for infiltration/exfiltration
k_{ij}	intrinsic permeability of a porous medium
l_{top}	thickness of a low-conductivity layer (aquitard) overlying the aquifer unit
ls	subscript indicating the linear source number s
l_s	geometry of the line defining the position of the linear source number s
m	Genuchten's soil parameter
n	effective porosity of the solid matrix, number of basis functions, Genuchten's soil parameter
n_{eff}	effective porosity
n_x	x component of a unit normal vector of the boundary
n_y	y component of a unit normal vector of the boundary
p	macroscopic pressure, average pore pressure within the REV
ps	subscript indicating the point source number s
q	unit volume flux
q_i	specific discharge (Darcy velocity)
q_i^{leak}	unit discharge (discharge per unit area) from the i-th leakage source
q_z^{bot}	rate of leakage through the base of the aquifer
q_z^{top}	rate of leakage through the top aquifer boundary, rate at which water enters storage above the water table of a phreatic aquifer
q_{G1}	potential flux at the soil surface
r	subscript which stands for 'relative'
s	subscript which stands for' 'solid' – a natural coordinate tracing a pathline
t	time
t_c	travel (concentration) time from a cell to the outlet,
t_k	k-th time level
ν	trial or weighting functions
$\bar{\nu}$	trial or weighting functions along the domain boundary
$\mathbf{x}_0$	centre of the REV
$\mathbf{x}$	point within the REV
x	Cartesian coordinate, global Cartesian coordinate
x_i	Cartesian coordinate in i-th direction
x_s	x coordinate of the source number s
y	Cartesian coordinate, a global Cartesian coordinate
y_s	y coordinate of the source number s
z	vertical Cartesian coordinate pointing upwards, vertical coordinate pointing downwards
A	coefficient
A_{as}	coefficient for the areal source number s
A_j	incremental areas for the time-area diagram
A_{ls}	coefficient for the linear source number s
A_{ps}	coefficient for the point source number s
B	coefficient, saturated aquifer thickness
B_{as}	coefficient for the areal source number s
B_{ls}	coefficient for the linear source number s
B_{ps}	coefficient for the point source number s

C	coefficient, the soil water capacity
C	concentration of a contaminant within the control volume W
C_{as}	concentration of a contaminant at the areal source number s
C^{bot}	concentration of a contaminant in the recharge through the aquifer base
C_{ls}	concentration of a contaminant at the linear source number s
C_{ps}	concentration of a contaminant at the point source number s
C_{sr}	surface runoff coefficient
C_G	concentration of a contaminant in groundwater along the model boundary
D	coefficient, diameter
D_3	segment based propagation in pathlines tracking
D_n	node based propagation in pathlines tracking
ET	potential evaporation
ET_0	reference potential evaporation
ET_p	potential evapotranspiration
H	potentiometric head (hydraulic head), groundwater level
H_i	the weighting coefficient at a Gauss's point x_i
H_{min}	the lowest level of the water table that still influences the recharge from a sewer
H_s	a representative hydraulic head for a sewer
H^*	Hubbert's fluid potential, the potentiometric head
$\bar{H}$	hydraulic head prescribed at the boundary
Inflow	the volume of water that enters the control volume
J	Jacobian matrix
K_{ij}	hydraulic conductivity tensor, i-th coefficients of the j-th equation
K_{top}	hydraulic conductivity of a low-conductivity layer (aquitard)
K_s	saturated hydraulic conductivity
K_x	saturated hydraulic conductivity in the x direction
K_y	saturated hydraulic conductivity in the y direction
K_z	saturated hydraulic conductivity in the vertical direction.
L	general form of the basic PDE, length
M	thickness of the aquifer unit, molecular weight of water
N	total number of computational nodes
N_1, N_2, N_n	basis functions in the form of polynomials
N_a	number of areal sources
N_i	rate of accumulation of water entering from external sources, polynomial number i
N_j	polynomial number j, the number of the computational node
N_l	number of linear sources
N_p	number of point sources
N'	transpose of N
Outflow	the volume that leaves the control volume
P	net precipitation
PAT	a 'pattern' function which describes how a quantity changes with time
P_0	reference precipitation
P_{eff}	the effective precipitation

Q	runoff at the outlet
Q_{as}	volumetric recharge rate from the areal source number s
Q_{ls}	volumetric recharge rate from the linear source number s
Q_{ps}	volumetric recharge rate from the point source number s
R	the universal gas constant
R_1	excess rainfall at the first time step
R_{off}	direct runoff
RH	relative humidity
S	aquifer storativity
Storage	the change in the volume stored
S_0	specific storage or specific storativity of a porous medium
S_e	relative saturation
S_{ij}	i-th coefficients of the j-th equation
S_s	specific storativity
S_t	terrain slope
S_y	effective porosity (specific yield) as related to the water table
T	air temperature
T_x	aquifer transmissivity in the x direction
T_y	aquifer transmissivity in the y direction
U_j	values of the unit hydrograph
V_i	fluid velocity component in i-th direction averaged over the volume of fluid
V_t	travel velocity
V_x	the fluid velocity component in x direction
V_y	the fluid velocity component in y direction
V_z	the fluid velocity component in z direction
W	width
W_2^2	class of functions
W_{max}	maximum water content
W_r	residual water content
Z_{bot}	elevation of the base of the aquifer unit
Z_{ter}	terrain level (land elevation)
Z_{top}	elevation of the top of the aquifer unit
α	Genuchten's soil parameter, compressibility of the porous skeleton
β	fluid compressibility
d_{as}	Diraq delta function which defines the position of the areal source number s
d_{ls}	Diraq delta function which defines the position of the linear source number s
d_{ps}	Diraq delta function which defines the position of the point source number s
e	macroscopic volumetric strain of the solid matrix, weighting factor
μ	fluid dynamic viscosity
ν_i	velocity of a solid boundary
r	fluid density, water density
θ	coefficient defining the collocation point, water content
θ_0	water content of unsaturated soil

x	local Cartesian coordinate
x_i	Gauss's point
h	local Cartesian coordinate
Dt	computational time step, time interval
Δz	spatial step
Δt_k	k-th time step
Γ	boundary
G_1	upper boundary of the UNSAT model domain
Γ_H	boundary with prescribed hydraulic head
Γ_q	boundary with prescribed unit discharge
W	plan area of a control volume, simulation domain
ψ	general flow property
∇	gradient

Glossary

Abstraction   Removal of water from any source, either permanently or temporarily.

Algorithm   An effective method for solving a problem expressed as a finite sequence of instructions.

Anisotropy   The property of being anisotropic; having a different value when measured in different directions.

Aquifer   Geological formation capable of storing, transmitting and yielding exploitable quantities of water.

Aquifer recharge; *syn.* groundwater recharge   Process by which water is added from outside to the zone of saturation of an aquifer, either directly into a formation, or indirectly by way of another formation.

Aquitard; *syn.* semi-confining bed   Geological formation of low hydraulic conductivity which transmits water at a very slow rate.

Artificial recharge   Augmentation of the natural replenishment of groundwater in aquifers or groundwater reservoirs by supply of water through wells, through spreading or by changing natural conditions.

Baseflow; *syn.* base runoff   Discharge entering a stream channel mainly from groundwater, but also from lakes and glaciers, during long periods when no precipitation or snowmelt occurs.

Capillary tension   The measure of the forces of attraction between a water molecule and the soil grain surface, representing the sum of the adhesive and cohesive forces or the capillary forces.

Capillary force   The force of attraction between a water molecule and the soil grain surface, representing the sum of the adhesive and cohesive forces.

Cartesian coordinate system   A coordinate system which specifies each point uniquely in a space by signed distances to three mutually perpendicular planes.

Catchment   The surface area bounded by topographical features, which drains to a single downstream location.

Compressibility   A measure of the relative volume change of a fluid or solid as a response to a pressure change.

Computer code   The set of commands used to solve a mathematical model on a computer.

Conceptual model   A general description of an object, process or a system, which may only be drawn on paper, described in words, or imagined.

Confined aquifer   Aquifer overlain and underlain by an impervious or almost impervious formation.

Database   An organized collection of data.

Delineation   The procedure of dividing the catchment area into sub-areas or sub-catchments.

Drainage   Removal of surface water or groundwater from a given area by gravity or by pumping.

Discharge   Volume of liquid per unit time that passes through a cross-section of a tube or a stream.

Effective porosity   Amount of interconnected pore space available for fluid transmission. It is expressed as the ratio of the volume of the interconnecting interstices to the gross volume of the porous medium, inclusive of voids.

Exfiltration   Flow of water out of a medium or an object such as pipe.

Evapotranspiration   Quantity of water transferred from the soil to the atmosphere by evaporation and plant transpiration.

Flow velocity   Vector indicating the speed and direction, at a point, of a moving liquid, e.g. water.

Flow rate; *syn.* discharge   Volume of liquid per unit time that passes through a cross-section of a tube or a stream.

Free surface flow   Flowing water having its surface exposed to the atmosphere.

Groundwater   Subsurface water occupying the saturated zone.

Graphical object   (in UGROW) Point, polygon or a set of points and polygons.

Groundwater flow   Movement of water in an aquifer.

Groundwater level   Elevation, at a certain location and time, of the water table or piezometric surface of an aquifer.

Groundwater recharge; *syn.* aquifer recharge   Process by which water is added from outside to the zone of saturation of an aquifer, either directly into a formation, or indirectly by way of another formation.

Groundwater storage   Quantity of water in the saturated zone of an aquifer.

Hydraulic head   Elevation to which water will rise in a piezometer connected to a point in an aquifer. Sum of the elevation and the pressure head in a liquid, expressed in units of height.

Hydraulic conductivity   Property of a saturated porous medium which determines the relationship, called Darcy's law, between the specific discharge and the hydraulic gradient causing it.

Hydraulic gradient   (in porous media) Measure of the decrease in head per unit distance in the direction of flow.

Hydrogeological boundary   Lateral discontinuity in geological material, marking the transition from the permeable material of an aquifer to a material of significantly different hydrogeological properties.

Hydrogeology   That branch of geology which deals with groundwater, in particular, its occurrence.

Hydrograph   A graphical representation of stage, that is, water depths above some datum, or discharge as a function of time.

Hydrological cycle; *syn.* water cycle   Succession of stages through which water passes from the atmosphere to the earth and returns to the atmosphere: evaporation from the land or sea or inland water, condensation to form clouds, precipitation, interception, infiltration, percolation, runoff and accumulation in the soil or in bodies of water and re-evaporation.

Impervious   Having a texture that does not permit water to move through it perceptibly under static pressure ordinarily found in subsurface water.

Impervious boundary; *syn.* no-flow boundary   Boundary of a flow domain through which no flow can take place because of greatly reduced permeability on the other side of the boundary.

Incompressible fluid   A fluid with negligible compressibility i.e. which does not change volume as a response to a pressure change.

Infiltration   Flow of water through the soil surface into a porous medium, such as the soil, or from the soil into a drainage pipe.

Infiltration capacity   Maximum rate at which water can be absorbed by a given soil per unit area under given conditions.

Integrated management   A planning and operational process in which interested parties, stakeholders and regulators reach general agreement on the best mix of conservation, sustainable resource use and economic development and diversification.

Isochrone map   Map or chart of a drainage basin in which a series of lines (isochrones) gives the times of travel of water originating on each isochrone to reach the outlet of the basin.

Leakage   The flow of water into or from an aquifer through an underlying or overlying semi-pervious layer, or any other source of water.

Mathematical model   A model which uses mathematical language, usually governing equations, to describe a system.

Matrix   A rectangular array of numbers set out by rows and columns.

Model   Representation in any form of an object, process or system.

Model calibration   Adjustment of the parameters of a model, either on the basis of physical considerations or by mathematical optimization, so that the agreement between the observed data and estimated output of the model is as good as possible.

Numerical model   An approximate solution of a mathematical model.

Optimizing   Choosing the best element from some set of available alternatives.

Phreatic surface; *syn.* phreatic level   A natural groundwater table (level).

Piezometer   A well, totally encased except at its lowest end, used to measure the hydraulic head at that point.

Point source   A stationary location or fixed facility from which pollutants are discharged; any single identifiable source of pollution; e.g. a pipe, ditch, ship, ore pit, factory smokestack.

Pollutants   Any substance introduced into the environment that adversely affects the usefulness of a resource or the health of humans, animals or ecosystems.

Pollution   Addition of pollutant to water.

Porosity   Ratio of the volume of the interstices in a given sample of a porous medium, e.g. soil, to the gross volume of the porous medium, inclusive of voids.

Porous matrix   Permeable medium containing connected interstices which may be considered as a continuous medium with respect to its hydraulic properties.

Porous medium   Permeable medium containing connected interstices which may be considered as a continuous medium with respect to its hydraulic properties.

Precipitation   Liquid or solid products of the condensation of water vapour falling from clouds or deposited from air on the ground. Amount of precipitation on a unit of horizontal surface per unit time.

Pressure head   Height of a column of static water that can be supported by the static pressure at a point.

Rainfall intensity   Rate at which rainfall occurs, expressed in units of depth per unit of time.

Rainfall intensity pattern   Distribution of rainfall rate, in time, during a storm.

Residual water content   Water content which remains in a sample of porous medium after it has been exposed to extremely low pressure for very long time.

Runoff   That part of precipitation that appears as streamflow.

Saturated zone   Part of the water-bearing material in which all voids, large and small, are filled with water.

Soil moisture profile   Curve representing the variation of soil moisture versus depth.

Sensitivity The relationship of the change of a response to the corresponding change of a stimulus, or the value of the stimulus required to produce a response exceeding, by a specified amount, the response already present due to other causes.

Sewage; *syn*, wastewater   The waste and wastewater produced by residential and commercial sources and discharged into sewers.

Sewer   An underground system of conduits (pipes and/or tunnels) that collect and transport wastewaters and/or runoff; gravity sewers carry free-flowing water and wastes; pressurized sewers carry pumped wastewaters under pressure.

Sewerage   The entire system of sewage collection, treatment and disposal.

Stormwater   Runoff from buildings and land surfaces resulting from storm precipitation.

Septic system   An on-site system designed to treat and dispose of domestic sewage. A typical septic system consists of a tank that receives waste from a residence or business and a system of tile lines or a pit for disposal of the liquid effluent (sludge) that remains after decomposition of the solids by bacteria in the tank, which must be pumped out periodically.

Specific storage   Volume of water released from or taken into storage per unit volume of the aquifer per unit change in head.

Subsidence   Lowering in elevation of a considerable area of land surface, due to the removal of liquid or solid underlying material or removal of soluble material by means of water.

Sustainable living   The potential for long-term maintenance of well-being, which in turn depends on the well-being of the natural world and the responsible use of natural resources.

Tensor   Geometric entity which extend the notion of scalars, geometric vectors and matrices.

Transient flow; *syn*. unsteady flow   The condition whereby, at any point in the flow field, either the magnitude or direction of the flow velocity varies with time.

Triangulation   The process of generating a set of triangles that connect all terrain points.

Unconfined aquifer   Aquifer containing unconfined groundwater having a water table and an unsaturated zone.

Unsaturated zone; *syn*. unsaturated soil   Subsurface zone above the water table in which the spaces between particles are filled with air and water, and the water pressure is less than atmospheric.

Unsteady flow   Flow in which the velocity changes in magnitude or direction with respect to time.

Urban drainage   A system of conveyance and storage elements serving to drain urban areas.

Urban water management   The process of planning, designing, building, operating and restoring urban drainage systems (an inter-disciplinary subject involving several professional and trade skills).

Urban water cycle   A water cycle including all the components of the natural water cycle with the addition of urban flows from water services, such as the provision of potable water and collection and treatment of wastewater and stormwater.

Urbanization   The trend seen in many urban centres in which populations increase and density of inhabitation also increases.

User interface   (of a computer program) The graphical and textual information the computer code presents to the user.

Vector A quantity that can be resolved into components.

Vadose zone; *syn.* unsaturated zone, zone of aeration   Subsurface zone above the water table in which the spaces between particles are filled with air and water, and the water pressure is less than atmospheric.

Water content   The volume of water contained in a sample of porous medium per unit volume of the sample.

Water management   Planned development, distribution and use of water resources.

Water quality   physical, chemical, biological and organoleptic (taste-related) properties of water.

Wastewater   Water containing waste, i.e. liquid or solid matter discharged as useless from a manufacturing process.

Wetting front   Air/water interface in the process of water infiltrating into a soil.

List of Contributors

- Ken W. F. Howard, Department of Physical and Environmental Sciences, University of Toronto, Toronto, Canada
- Dubravka Pokrajac, School of Engineering, University of Aberdeen, Aberdeen, United Kingdom
- Christina Schrage, Project Manager Geo Ecology, Karlsruhe, Germany
- Miloš Stanić, Institute of Hydraulic Engineering, Faculty of Civil Engineering, Belgrade, Serbia
- John H. Tellam, School of Geography, Earth and Environmental Sciences, University of Birmingham, Birmingham, United Kingdom
- Leif Wolf, Institute for Applied Geosciences, University of Karlsruhe, Karlsruhe, Germany

Preface

UGROW as an IHP-VI Component

Ensuring healthy and sustainable living conditions in intensively populated areas has emerged as a major global challenge, and the provision of safe and sustainable water supplies for drinking and sanitation is central to this issue. Important management decisions will have to be taken and it is essential that these acknowledge the entire urban water cycle and the complex interactions that take place between groundwater, surface water and the complex network of water services, including sewers and pressurized water supply systems.

Historically, the vital role that groundwater plays in the urban water cycle has been severely neglected. To a certain extent this reflects an 'out of sight, out of mind' mentality, resulting in ignorance of subsurface water movements. However, neglect has also arisen because groundwater and surface water systems are spatially distinct, and in terms of water flow velocities, they operate on totally different timescales. Reasons aside, the unfortunate consequence is that tools for urban water management rarely, if ever, incorporate an adequate understanding of urban aquifers and the role of groundwater – either during the analysis stage or, just as importantly, during the subsequent decision-making process. These attitudes must change and time is of the essence. The need to prioritize holistic management of the urban water cycle is gaining increasing recognition worldwide. In turn, practical, soundly developed urban water system modelling tools are essential if the goal of urban sustainability is ever to be achieved.

In the face of such challenges, *UGROW* (Urban**GRO**und**W**ater) represents one of the most advanced urban water management tools produced to date. Developed under the sixth phase of UNESCO's International Hydrological Programme (IHP-IV), *UGROW* fully integrates all urban water system components including groundwater. Its principal aims are to raise awareness of the interaction between urban water system components, to support management decision-making, and to solve a wide range of urban water problems. The model has a sound scientific basis, is computationally efficient and is supported by excellent graphics. It has also been tested and refined under a range of demanding urban conditions.

Ken W. F. Howard and Dubravka Pokrajac

Chapter 1

Challenges in urban groundwater modelling as an introduction to *UGROW*

Ken W.F. Howard[1] and John H. Tellam[2]

[1]*Department of Physical and Environmental Sciences, University of Toronto, Toronto, Canada*
[2]*School of Geography, Earth and Environmental Sciences, University of Birmingham, Birmingham, United Kingdom*

1.1 THE MANAGEMENT OF URBAN GROUNDWATER

The world's population is increasing at an alarming rate, with much of this growth occurring in urban areas (Figure 1.1) (United Nations, 2005). Between 1990 and the turn of the twenty-first century, the global population grew by 15% (from 5.3 to 6.1 billion) while the population of urban areas increased by 24% to almost 3 billion, a number that is currently increasing at the rate of almost 200,000 per day. By 2010 over half the world's population will live in urban areas, and by 2030 the number of urban dwellers is expected to reach almost 5 billion, or 60% of the projected global population of 8.2 billion.

Urban areas are the economic power engines of the world, but their long-term sustainability relies heavily on the provision of adequate water supplies. In response, the need for holistic management of the entire urban water cycle has emerged as a major priority amongst the more proactive cities and reveals a growing awareness of the interactive nature of groundwater and surface water in urbanized catchments and the need to optimize water use through an integrated approach. Unfortunately, the critical role groundwater plays in the urban water cycle continues to be neglected in many parts of the world (Howard and Gelo, 2002) and has not been adequately accommodated within the principles of IWRM (Integrated Water Resources Management) (Van Hofwegen and Jaspers, 1999; Global Water Partnership, 2000, 2002). To some extent, this problem arises

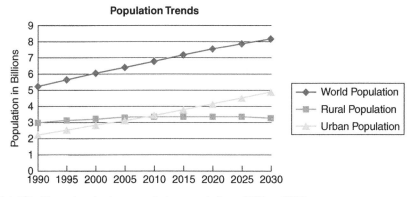

Figure 1.1 World, rural and urban population trends from 1990 to 2030

Source: United Nations, 2005

because groundwater and surface water operate on distinctly different spatial and time scales, and approaches to their management, including the development of modelling tools, have evolved independently. This must change if the world's rapidly growing cities are to entertain any hope of achieving long-term sustainability. Groundwater needs to become fully integrated within IWRM, and urban water system modelling tools that can seamlessly incorporate all components of the urban water cycle are required.

Interest in the relationship between urban development and water has had a long history, instigated locally at different times according to local urban development. For example, in the mid 1900s, accelerating urban growth following the Second World War, notably in Europe and North America, triggered a wide range of serious hydrological problems. Most were related to a rapid rise in the expanse of impervious surface, which led to sharp increases in the frequency and intensity of urban flooding. Within a few years the discipline of 'urban hydrology' became firmly established and attracted well-funded researchers from a broad range of disciplines.

Urban groundwater issues have taken longer to emerge and, as a consequence, the science of urban groundwater is comparatively young. To date, remediation and problem resolution have taken priority over much-needed proactive measures such as aquifer management and groundwater protection. Nevertheless, significant progress has been made on a number of key problems and a wealth of knowledge has been accumulated (Howard and Israfilov, 2002; Lerner, 2003; Tellam et al., 2006). Broadly, and as reviewed in detail in Section 1.2, these issues relate to the impacts of urbanization on either the *quality* or *quantity* of the groundwater resource. As indicated in Section 1.2.5, such issues pose important challenges in the development of simulation models suited to the urban groundwater environment.

Developments in our knowledge and understanding of urban groundwater during the past twenty-five years have been matched by advances in our ability to simulate aquifer behaviour and contaminated groundwater using numerical modelling techniques. As discussed in Section 1.3, most advances in the field of numerical modelling have been generic in the sense that the vast majority relate to the simulation of basic flow and transport processes, typically within the zone of saturation. Very few have focused on types of hydrogeological conditions unique to urban areas. This is beginning to change. In the past decade, considerably more attention has been paid to the development of models able to incorporate urban features such as leaky pipes and sewers. *UGROW* (**UrbanGROundWater**) represents one of the most advanced, fully integrated, urban groundwater models produced to date.

1.2 WHAT HYDROGEOLOGICAL CHARACTERISTICS ARE UNIQUE TO URBAN GROUNDWATER SYSTEMS?

The development of *UGROW* was inspired by the premise that urban groundwater systems have unique characteristics, which make the development of a specialized groundwater modelling tool necessary. These characteristics are explored here.

Examination of the results of many urban groundwater studies (e.g. those in Chilton et al. (1997, 1999); Howard and Israfilov (2002); Tellam et al. (2006)) suggests that few, if any, basic processes are truly unique to urban systems. What sets urban hydrogeology apart from the rest of hydrogeology is the frequency of occurrence of certain 'elements' that have an effect on either the groundwater flow system or the chemistry of the groundwater. These 'elements' are generally associated with residential, transport

Table 1.1 Elements of greater importance in urban areas due to their more frequent occurrence

Elements	Effects/Comments
Geology	
'made ground'	change in hydraulic properties affecting recharge; supply of solutes
foundations and cutoff walls	change in flow patterns
induced landslides	change in hydraulic properties
abstraction-induced subsidence	change in surface hydrology; inducing of pipeline leakage; change in aquifer system properties
Aquifer Recharge	
paved cover	increased runoff; reduced infiltration; reduced evapotranspiration; enhanced fingering/funnelling: locally there may be increased recharge if drains associated with paved areas are connected to soakaways rather than storm sewers
interception by buildings & roads	reduced recharge unless soakaways present
pipeline leakages	increased recharge
sewer leakage	often relatively small increase in recharge
industrial discharges	often relatively small increase in recharge
urban micro-climates	changed evapo(transpi)ration; rainfall
groundwater abstraction	increased recharge through increased vertical & horizontal gradients
artificial recharge	increased recharge
Aquifer Discharge	
abstraction	complex, sometimes rapidly changing flow patterns; low water levels
passive drainage	drain systems diverting runoff to storm sewers
evapotranspiration	dependent on water levels in aquifer; may be limited because limited vegetation cover
discharge to surface waters	change of flow regime; possible change in surface water capacity for diluting discharging groundwaters
tunnels	change in discharge location
Groundwater Chemistry	
atmospheric precipitation	acid precipitation; construction wash-out
runoff	often good quality inorganically, except where de-icing salts used, but can have poorer organic and biological quality
pipeline leakages	good quality water main leakages; chlorination by-products; chemical pipeline leakages
sewer leakages	poor quality inorganically, organically and microbiologically
releases from made ground	long-term source of pollutants
surface water infiltration	quality dependent on surface water quality and interactions with river bed sediments
surface water exfiltration	no direct effect, but groundwater may affect surface water quality, dependent on river bed sediment compositions
industrial discharges	wide range of quality; short-term to long-term releases, plume fragments to large plumes
abstractions	encourages migration of plumes, including to depth; within-well and through-well mixing affects abstracted groundwater quality
total loading & attenuation capacity	may exceed total attenuating capacity of aquifer on regional or local scale; NAPLs important in latter context
'new' chemicals	new synthetic organics; manufactured nanoparticles; uncertain environmental behaviour
'old' chemicals	'obsolete' chemicals may still be present in aquifer
major changes in abstraction rate	change in groundwater level may change chemical processes, for instance, leading to change in redox conditions
mixing	in-well mixing especially; danger of mixing resulting in reactions with products of greater toxicity

and industrial activities, and are listed in Table 1.1. All are present to a lesser extent in non-urban areas, but can usually be ignored without significant loss of accuracy in any regional groundwater assessment. Likewise, there are certain elements in rural aquifers (e.g. trees and diffusely-spread pesticides) which, although sometimes present in urban aquifers, are not normally important elements of the latter systems.

The issues and implications for flow and chemical transport that arise from the high frequency of occurrence of elements listed in Table 1.1 are examined below. The discussion is divided loosely into four themes (Table 1.1), which broadly reflect some of the main aspects of model design: geology, aquifer recharge, aquifer discharge and groundwater chemistry. There is inevitably some overlap in subject matter between the different themes and not all aspects of each theme are discussed, as many are not specific to urban systems.

1.2.1 Geology

The distribution of the main geological units underlying a city are unaffected by urban development. However, there are a few situations where urban development affects the disposition of shallow subsurface material.

- *Made ground:* These deposits can cover a significant proportion of a city area, affecting both groundwater flow and groundwater chemistry. 'Made ground' or 'fill' refers to anthropogenic material, for example, building, industrial or domestic waste (Rosenbaum et al., 2003) used to infill depressions and provide a level surface for construction. For the purpose of this discussion, the term 'made ground' excludes surface coverings such as paved surfaces and buildings, which are discussed in Section 1.2.2 (aquifer recharge).

  Made ground is often heterogeneous in composition and, as a result, its hydraulic and chemical properties often display similar heterogeneity. Some components of made ground are chemically inert, but others, for example, plaster or leachable industrial or domestic wastes, can be very reactive.

  Made ground can also be present as infill material in trenches cut into non-anthropogenic or other made ground deposits (Brassington, 1991; Heathcote et al., 2003). Such trenches can have significantly greater hydraulic conductivity than the surrounding material, and can also be associated with potential pollution sources, for example, sewers.

  Usually little is known of the composition or properties of made ground at any given site. Characterization is difficult given the degree of heterogeneity, and geophysical methods of investigation which might otherwise be of considerable use are often rendered impracticable because of paved surfaces, the presence of services and metal objects within the made ground. The same problems can also reduce the utility of geophysical methods for investigating the aquifer below made ground.

  Sometimes, however, site investigation borehole data may be available from previous construction work. Where the made ground has low permeability, recharge may be limited. In other systems, perching may occur and can result in the redistribution of recharge and pollutant fluxes. If funnelling of water via permeable zones occurs, the residence time in the unsaturated zone will be decreased, as will contact with the potentially attenuating aquifer materials. Often the role of made

Figure 1.2 A recent major urban landslide in Baku, Azerbaijan (site **X**), was directly associated with high groundwater levels, caused by a combination of heavy rain and leaking water mains (see also colour plate 1)

Source: The authors

ground is uncertain, and modelling studies have to explore various conceptualizations (e.g. Heathcote et al., 2003). As a complicating factor, rising groundwater levels due to falling abstraction or change in climate may result in a change in the redox status of shallow contaminated made ground, either enhancing or inhibiting pollutant breakdown.

- *Foundations and cutoff walls:* In shallow aquifers, foundations can significantly affect groundwater flow patterns, producing both damming and funnelling features. Foundations may also have an effect on water quality, one example being the use of chemically reactive grouting (Eiswirth et al., 1999). In some cases, slurry trenches or other cutoff walls may affect shallow groundwater flow.
- *Induced landslides:* Landslides can be initiated within urban areas as a result of anthropogenic activities, especially rising water levels (e.g. resulting from mains leakage) (Figure 1.2), or loading by buildings (e.g. Alekperov et al., 2006). The slipped material, may also contain entrained building rubble, thus locally affecting recharge and water quality.
- *Subsidence:* Intensive urban abstraction in areas underlain by sand/clay sequences can result in significant ground subsidence, as has occurred, for example, in Mexico City and Bangkok, and which started to occur in Venice. This may have a significant effect on urban surface drainage, induce leakage from fractured pipelines and change the hydraulic properties of the aquifer system.

1.2.2 Aquifer recharge

Urban development often has a profound effect on aquifer recharge. Contributory factors include the degree to which the ground surface covering is altered, the high density of potable and wastewater systems, the high densities of intentional and unintentional discharge of waters to the subsurface, and the inducement of infiltration brought about by a high concentration of abstraction wells. Given the complex land cover within a city (e.g. Figures 1.3 and 1.4), recharge distributions tend to be extremely heterogeneous. Made ground effects are briefly considered above; here the discussion focuses on paving, buildings and discharge of waters.

- *Paving:* Although vegetated surfaces still comprise a significant proportion of the total area in many cities (e.g. Figures 1.3 and 1.4), paving of urban catchments is very much more extensive than in rural catchments. This encourages surface runoff and reduces direct infiltration. However, few paved systems are totally impervious with infiltration through defects being encouraged in many cases by surface water ponding (e.g. Cedergren, 1989). This may result in rapid funnelling of water passing through the unsaturated zone (e.g. Kung, 1990), with almost no evapotranspiration and possibly only limited natural attenuation (Thomas and Tellam, 2006a). As infiltration in this case will be almost evapotranspiration-free, recharge will show much less variation seasonally than is the case where vegetated surfaces are present. Rapid funnelling of water may also result in perching, either above low permeability layers, or above coarse units with high air entry pressures; the latter will encourage fingering (Glass et al., 1988).

  In some urban areas, paved systems have been engineered to have high permeability (e.g. Martin et al., 2001) to increase infiltration and reduce the 'flashiness' of runoff. Occasionally, but less intentionally, other cover material encourages recharge: an obvious example is the gravel bed often used for railway tracks, or soakaways associated with roads. Such cover materials may also be responsible for the rapid entry into the subsurface of pesticides and oils from maintenance work and vehicle discharges, or from tanker accidents (e.g. Atkinson and Smith, 1974; Lacey and Cole, 2003; Atkinson, 2003).

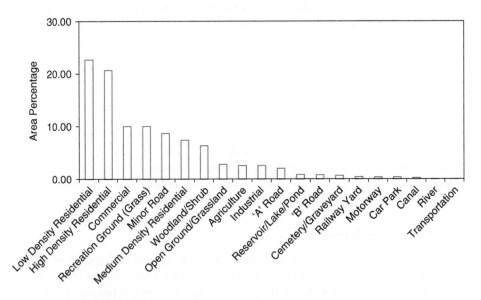

Figure 1.3 The proportions of land cover in the unconfined portion of the Birmingham urban sandstone aquifer, UK

Source: Thomas and Tellam, 2006a

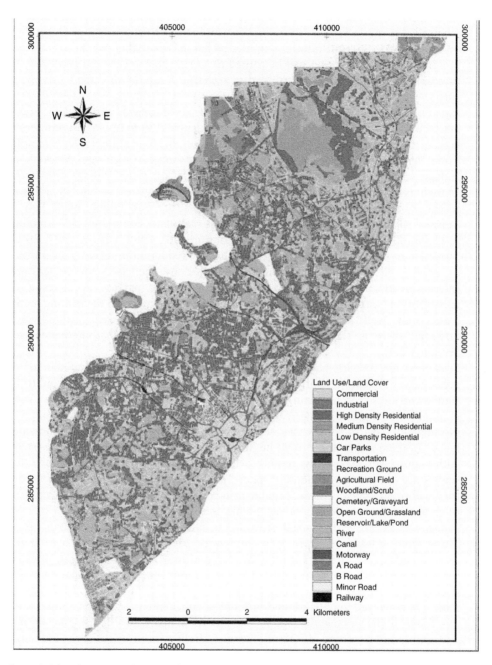

Land Use/Land Cover
Commercial
Industrial
High Density Residential
Medium Density Residential
Low Density Residential
Car Parks
Transportation
Recreation Ground
Agricultural Field
Woodland/Scrub
Cemetery/Graveyard
Open Ground/Grassland
Reservoir/Lake/Pond
River
Canal
Motorway
A Road
B Road
Minor Road
Railway

Figure 1.4 Land cover on the unconfined sandstone aquifer underlying Birmingham (see also colour plate 2)

Source: After Thomas and Tellam, 2006b. This map is based in part on Ordnance Survey data © Crown copyright Ordnance Survey

- *Interception by buildings and roads:* Roof interception will, in some buildings, be channelled to grassed swales or soakaways, and thence to groundwater. Depending on the role of evapotranspiration, aquifer recharge is probably enhanced. In other buildings, roof interception will be routed to stormwater sewers and precipitation that would normally have contributed to recharge is rapidly directed to receiving surface water systems.
- *Piped water leakage:* Although the presence of paved areas is expected to reduce net recharge, an extra source of aquifer replenishment in many urban aquifers is leakage from piped water systems (mains leakage). In some aquifers, for example, Lima, Peru, (Lerner, 1986) this may be the main source of recharge. Amounts of leakage vary considerably. Puri (pers. comm. 2004) estimates leakage rates to be as high as 90% in some rural towns in developing countries: in the UK, rates in recent years have been up to 30%, although these are now falling, while Van de Ven and Rijsberman (1999) estimate a leakage rate of approximately 5% for cities in Holland. In the UK, several studies (Rushton et al., 1988; Lerner, 1997, 2002, 2003; Knipe et al., 1993) have indicated that the reduction in recharge as a result of paving has been largely compensated for by increased mains leakage. Internationally, recharge in urban areas appears to increase as a result of urbanization (see summary by Foster et al., 1999) The quality of the leaking mains water may well be much better than that of the *in situ* groundwater, however, there is the possibility of the production of trihalomethanes (see Section 1.2.4).
- *Sewer leakage:* As with mains water, sewer systems are densely distributed in urban areas. Sewer systems also often leak, but being unpressurized, the rates tend to be significantly less than for mains leakage (Misstear et al., 1996). For example, Yang et al. (1999) estimated sewer leakage to be supplying <10% of total recharge in Nottingham, UK. In some cases, sewer elevations will be below the groundwater level; in these cases there will be infiltration from the groundwater system. Although, sewer leakage represents a line source of contamination, individual sewers are likely to be so densely spaced that, at common scales of observation, they will often appear diffuse. Similar principles often apply to other line sources of contamination, such as de-icing chemicals applied to roads.
- *Industrial discharge:* Industrial sites are usually concentrated in urban areas. Waste from industrial sites may be discharged to surface water systems, to the ground surface, or directly into the aquifer, and may be legal or illegal, in other words, controlled or uncontrolled. In many countries, legislation has changed over time, and what was previously legal is now illegal. In some cases, contaminants are conveyed via abandoned boreholes, either intentionally or unintentionally. Some discharges may be regular and recorded; many may be irregular in time and space, and unrecorded. Often, chemical discharges in urban areas are small and transitory – spillage from a container, for example. Sometimes, sources are short-lived because they are discovered relatively quickly through the auditing of storage-container fluid balances. Other sources are longer-lived because they remain undiscovered for many years or are permitted by the regulatory authority.
- *Urban microclimates and soil heating:* Because of the heat they store and discharge, cities change the local climate and warm the subsurface, often reversing the geothermal gradient. This can affect recharge rates, changing both precipitation rates and evapo(transpi)ration (e.g. Grimmond and Oke, 1999). Increasingly,

urban areas are using groundwater and heat exchange systems to control building temperatures throughout the year (e.g. Anon, 2002), further affecting flow systems and potentially affecting reaction rates.

- *Recharge via surface water:* Many cities are built on rivers. Recharge from urban rivers has been induced in many urban aquifers by overabstraction and in some cases by purpose-designed 'bank infiltration' schemes (Hiscock and Grischek, 2002). Influx is partly controlled by river bed deposits. The hydraulic properties of these may be modified by suspended solids contained in wastewater that has been discharged into the river upstream – anthropogenic colmation. Similar comments can be made about urban lakes and canals: canals in particular are common in urban areas (e.g. Birmingham, a major city in central UK, has approximately 180 km of canals). If an aquifer is located on the coast and is heavily pumped, sea water intrusion is likely, with the potential for destroying the local aquifer (e.g. Howard, 1988; Carlyle et al., 2004). In some cases, complex injection well schemes have been implemented to control the intrusion. Some coastal urban systems have been modified by the construction of barrages (e.g. Cardiff, UK), and these will have major effects on local groundwater flow systems (Heathcote et al., 2003). In addition, coastal aquifers, including those on estuaries, will be sensitive to rising seawater levels associated with climate change.

- *Changing water levels:* Falling water levels due to localized urban abstraction will increase head gradients across the unsaturated zone, thus increasing recharge, especially where shallow, lower permeability deposits, such as some types of made ground, are present. Conversely, recharge rates are likely to decrease if water levels are rising in cases where the city has recently moved to a remote water supply, or economic pressures have lowered industrial production rates.

- *Artificial recharge:* As urban water demand rises, more cities are considering various means of artificial recharge or 'recharge management' (e.g. Dillon and Pavelic, 1996; Chocat, 1997; Pitt et al., 1999). These include basin recharge, borehole injection including (artificial) aquifer storage and recovery (ASR) (Pyne, 2005; Jones et al., 1998), and, although often considered more as a means of reducing urban runoff, permeable pavements. Water sources include urban drainage waters such as stormwater runoff and brown/green roof rainfall harvesting. The recharged water is later recovered, either at the same location (e.g. in ASR schemes), or at some remote location. The former approach is appropriate where the 'grey' water is of adequate quality for local use (e.g. washing or cooling), whilst the latter takes advantage of the aquifer's natural capacity to attenuate various contaminants. In some cases, the grey water may actually be of better quality than the groundwater into which it is injected. Normally, however, the injected water is of poorer quality, especially in terms of particulates and microbes (e.g. Datry et al., 2006; Anders and Chrysikopoulos, 2005). For large schemes, the injection water is treated, if only to reduce the rate of clogging.

1.2.3 Aquifer discharge

Most discharge boundaries in urban aquifers are of the same types as those found in non-urban aquifers. However, the greater density of abstraction wells and the greater degree to which surface drainage systems are modified in urban aquifers both influence the character of urban aquifer discharge processes.

- *Abstraction:* The density of abstraction points is a major characteristic of many, though not all, urban aquifers, particularly in the early development of a city before the dangers of polluted water are considered and large-scale water importation occurs (Morris et al., 1997). Abstraction patterns in many cities have been, and in many cases still are, controlled by non-hydrogeological factors, including transport routes, historical land ownership and pre-existing water distribution networks. As a result, pumping frequently causes interference, and flow patterns can be complex, especially where pumping rates vary diurnally or seasonally, and/or the pumping is unregulated. The flow system can be very different from that occurring naturally, especially where over-abstraction has occurred (e.g. Knipe et al., 1993). Internal boundaries such as faults may be more apparent in urban aquifers because of the disruption of flow caused by heavy abstraction (e.g. Seymour et al., 2006). Construction will often require water control, and this again will affect the urban flow systems (e.g. Brassington, 1991; Preene and Brassington, 2003; Attanayake and Waterman, 2006). Generally, this will be a temporary effect, although in a rapidly expanding urban area many sites may be developed at approximately the same time. In a few cases, dewatering will need to be continuous, for example, in the case of some deep tunnels (Rushton et al., 1988). The complexity of pumping regimes and the lack of detailed records of pumping rates and times make the modelling of flows and solute transport very difficult in this environment. Historical records can sometimes be unearthed, but vital historical data are often permanently lost.
- *Passive drainage:* Transport routes often require drainage systems, as described above (Brassington, 1991). In some cases, this will result in the collection of runoff and discharge to soakaways, thus increasing recharge. However, in other cases the collected water will be transferred to the storm sewer system and mostly discharged outside the aquifer area.
- *Evapotranspiration:* Heavy abstraction will lead to falling water levels, and in some cases this in turn will result in reduced evapotranspiration losses (e.g. Khazai and Riggi, 1999).
- *Discharge to surface waters:* Urban water courses are usually highly modified as a result of the construction of drains, canalization and culverting (e.g. Petts et al., 2002; Bradford, 2004). Due to this, and because of discharges from sewerage and industrial works and modification of catchment runoff characteristics, hydrographs can be considerably modified compared with their natural state, often being more 'flashy'. This will affect the relationship between baseflow and the other hydrograph components, emphasizing the response time differences between surface water and groundwater systems.

1.2.4 Groundwater chemistry

Urban aquifers are prone to quality problems, not least because a city produces large amounts of waste, some of which inevitably finds its way into the groundwater system. Here, some of the main issues are briefly summarized.

- *Atmosphere and precipitation chemistry:* Atmospheric releases from power generation, industry and vehicles result in a range of reactions between the gaseous, aqueous and particulate phases in the atmosphere, and as a consequence, precipitation chemistry is changed. A common result is the production of 'acid rain', which falls within the city and therefore affects urban recharge quality; falls in neighbouring rural areas; or is transferred over greater distances. In general, because of the different rates of reaction within the air, HCl (often from coal combustion) is likely to be deposited close to the source. HNO_3 takes longer to form from NO_x, and therefore NO_3-rich acid rain will be deposited at greater distances from the source. H_2SO_4 is the slowest to form, and its effects will sometimes be felt up to many hundreds of kilometres from the source (e.g. Harrison and de Mora, 1996). Acid rain thus affects areas well beyond the urban perimeter, but the precise chemistry of the acid rain will change with distance: of course, acid rain in a given city may also partly derive from a neighbouring city. Other sources of pollutants in atmospheric inputs are construction dust and the usual marine and soil sources. Urban trees can collect polluted atmospheric moisture and convert it to recharge by stem flow or branch drip processes, this being more important in smog-prone cities which also contain significant numbers of trees.

• *Runoff quality:* The quality of runoff from paved surfaces is often relatively good, with low concentrations of major species, metals and organics, although in some cases, especially after dry periods, first flushes can have greater metal and organic concentrations. In cities where they are used, de-icing salts can give rise locally to brine concentration runoff from paved surfaces (Figure 1.5), and where this infiltrates, high recharge concentrations are expected. Even if much of the runoff is channelled to surface water discharge, some will infiltrate, and high steady-state groundwater chloride concentrations have been predicted for some cities, for

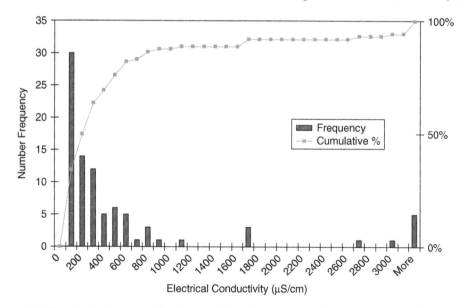

Figure 1.5 The electrical conductivity of road-drain gully-pot waters from the campus of the University of Birmingham, UK. Samples were collected mainly in winter, spring and autumn. Maximum conductivity recorded was in excess of 100,000 μS/cm.

Source: Harris, 2007

example, 400 mg/l for Toronto (Howard and Haynes, 1993; Howard and Beck, 1993). In addition to rock salt, urea and synthetic organic compounds are also used for de-icing purposes (e.g. Wejden and Ovstedal, 2006). Roof runoff quality will often not differ significantly from local rainfall quality, but organic matter entrained in the runoff may cause soakaways to become chemically reducing, and some roof runoff can yield significant amounts of heavy metals (into the hundreds of microgrammes per litre range (Harris, 2007)).

- *Pipeline leakage:* Although usually better in quality than the groundwater into which it is infiltrating – and therefore providing a certain amount of dilution – leaking mains water or other treated waters can contain residual Cl_2, which may react with groundwater or aquifer organic matter to produce trihalomethanes (Fram, 2003). Some contaminants derived from the dissolution of pipe materials may also be released in any leaking water, including Cu and Zn from metal pipes and phthalates from plastic pipes. Occasionally, other types of pipeline exist in urban areas, including oil and chemical transfer pipelines. High voltage electrical cabling is also sometimes laid within pipes containing organic fluids. These types of pipeline are significant as potential pollution sources.

- *Sewer leakage:* The quality of water infiltrating from sewers will be poor, with high levels of BOD, COD, Cl, N, microbes and, depending on the nature of local industrial wastes, metals and synthetic organics (e.g. Misstear et al., 1996; Barrett et al., 1999; Pedley and Howard, 1997; Powell et al., 2000; 2003; Wolf et al., 2004, 2006a; Cronin et al., 2006). Pharmaceutical and other medical compounds have also been found in measurable quantities in groundwaters in some locations (e.g. Scheytt et al., 1998; Held et al., 2006). Mains pipes are sometimes located above sewer pipes, and where leaks coincide or are compounded by flow along the pipeline trenches, mobility of the compounds released may be affected either adversely or advantageously. Some sewer systems are combined, taking both domestic 'foul' sewage and surface 'stormwater' runoff (e.g. Butler and Davies, 2000). In such systems, foul sewage is diluted, but there is also a greater possibility of sewer overflow during heavy storm events. Sewage infiltration will also occur from septic tanks, and where discharge is direct to the ground surface, for example, in the shanty townships attached to some of the largest cities in the world. Some components of contamination are relatively short-lived (Barrett et al., 1999), for example, human bacteria and viruses, the latter perhaps only surviving for a couple of years at most in groundwater systems (Sellwood, pers. comm., 2006; Joyce et al., 2007). Other sewage components are more recalcitrant, including N species and Cl. Urban nitrate concentrations are often equally as elevated as in rural settings (e.g. Lerner et al., 1999). Eventually these pollutants may achieve a quasi-steady-state concentration within the aquifer, but this can take decades, even if the relevant parts of the urban socio-economic-physico-chemical system remains stable.

- *Releases from made ground:* Some made ground is very reactive, particularly that derived from domestic waste, certain forms of industrial waste and some types of building waste. Given the volume of material involved, made ground may contribute solutes for long periods of time.

- *Surface water infiltration:* Where surface water bodies are polluted, there is a risk that pollution will enter the aquifer. There may, however, be some attenuation by the lake and river bed sediments (e.g. Smith, 2005).

- *Surface water exfiltration:* Baseflow discharge to urban rivers can be a useful means of diluting either polluted groundwater or polluted river water (e.g. Ellis and Rivett, 2006). The modification of urban surface drainage systems discussed in Section 1.2.3 may result in significant changes to the relationship between surface water components and baseflow: the greater 'flashiness' of river flows can mean reduced dilution of baseflow during low flow periods in some systems.

- *Industrial discharges:* Industrial discharges may vary widely in composition, many of which will not be found in other contexts. Short-lived sources will form plume fragments which will migrate through the aquifer, increasing in volume but reducing in concentration as they move. Discharges will often be small in quantity compared with other recharge sources, but in some cases will contribute significant amounts of contaminant. The presence of temporary, often relatively small discharges can result in a spotty distribution of industrial pollutants in wellwaters (e.g. Tellam and Thomas, 2002) – a distribution which rarely reaches a steady-state concentration, and is often difficult to relate with any precision to sources. In some cities, industrial sites are clumped, and hence pollution from discharges is also effectively clumped. However, petrol stations and dry-cleaning premises, two frequent sources of small-scale release of organic pollutants in urban areas, are often densely distributed across even residential areas. For example, in the unconfined part of the Birmingham aquifer (UK), there is around one petrol station for every $2\,km^2$ (Thomas and Tellam, 2006b).

- *Abstractions:* Abstraction from an aquifer can develop vertical head gradients locally, encouraging vertical penetration of pollutants (e.g. Taylor et al., 2003, 2006). If abstraction subsequently decreases, water levels will rise and re-saturate parts of the aquifer that were previously unsaturated. This may encourage further leaching of residual non-aqueous phase liquids, probably under anaerobic conditions rather than the aerobic conditions occurring previously. The spatial and temporal complexity of abstractions probably means that plumes within many urban aquifers are dispersed more quickly than they might be in less 'stirred' aquifers (Jones et al., 2002). This will tend to increase natural attenuation, but whether this is significant is uncertain. Finally, abstraction of polluted water will reduce the pollutant loads in the aquifer by what is in effect, pump and treat remediation (e.g. Lerner and Tellam, 1992; Rivett et al., 2005).

- *Total loading and attenuation capacity:* Where high concentration discharges occur over an extended period of time, the aquifer's attenuation capacity may be exceeded locally (e.g. Ford and Tellam, 1994). This effect may be at least partly counterbalanced by increased mixing (see *Abstractions* above). The general model of urban development proposed by Foster et al., (1999) provides an example: this model proposes that in unsewered systems there will be a gradual fall in redox potential within the urban system in response to high organic loadings, with a redox zonation developing within the aquifer. Another example is the acidification of a sandstone aquifer described by Ford et al. (1992) (see also Shepherd et al., 2006). At a local scale, the aquifer's attenuation capacity may be exceeded because of the presence of a non-aqueous phase liquid (NAPL) providing high concentrations of organic pollutants. It may take many years for a NAPL to be removed by dissolution, and during that time it will continue to act as a concentration boundary. The NAPL may represent quite a complex concentration boundary as its composition, flushing rate and even location may change with time. NAPLs are almost certainly common in many urban aquifers. In the UK, where chlorinated solvents have been used at a site, dissolved phase chlorinated solvents are generally present in site well waters (Rivett et al., 1990; Burston et al., 1993; Shepherd et al., 2006). In most cases, the presence of the dissolved phase implies the existence of the free phase, and in some cases the high dissolved phase concentration definitively shows this. However, many NAPLs will never be noticed in urban aquifers, and even if noticed will not necessarily be recorded.

- *'New' chemicals:* New chemicals are continually being produced by industry, and manufacturing sites are often located in urban areas. Most new chemicals are synthetic organics, but may in the future also include manufactured nanoparticles. Some of these may be toxic to humans and many are already known to be toxic to bacteria, including possibly those involved in degrading other pollutants (Anon, 2004).
- *'Old' chemicals:* As new chemicals replace older chemicals, the latter may still be worthy of consideration, as they may still be present within the groundwater system.
- *Major changes in abstraction rate:* If abstraction significantly changes, water levels will either rise or fall. This will often result in changing the redox status from aerobic to anaerobic, or vice versa, with consequent effects for chemical reactions. For example, aerobic degradation of petroleum hydrocarbons may no longer occur, or anaerobic degradation may cease for chlorinated solvents. Robins et al. (1997) report a case where falling water levels induced oxygen influx, resulting in sulfide oxidation and the production of very high sulfate, low pH waters.
- *Mixing:* Wells may penetrate one or more plumes, and abstraction will thus cause within-well mixing. This mixing will initiate reactions, and it is possible that harmful products may be formed, although we are not aware of documented examples. This mixing differs from the dispersive mixing described previously (see *Abstractions*): the mixed water is out of contact with the aquifer solid phase and is probably in contact with air, and reaction times are probably limited. In this context, mixing has greater similarity with discharge into surface waters, although here contact with organic matter will occur. The dilution effects of mains leakage have also previously been noted.

1.3 THE CHALLENGES FOR MODEL REPRESENTATION OF URBAN AQUIFERS

Many cities are developed beside rivers or on aquifers. Although such locations have obvious benefits, one drawback is that the waste produced by the high density human (and animal) occupation is likely to enter the main water source. Minimizing water quality deterioration whilst maximizing available water yield is the principal task of all water planners, but the water quality constraint is particularly acute for urban aquifers. The capability to deal with both flow and reactive solute transport should thus be included in any urban groundwater model used for urban water resource planning.

Ignoring special cases, such as seawater intrusion and ASR (Aquifer Storage and Recovery), urban aquifer systems have some specific and peculiarly problematic characteristics, including:

- large numbers of different recharge mechanisms, many of which, to date, lack tractable quantitative description (or even physical understanding),

- marked heterogeneity in recharge distribution in space and time,
- mismatch in response times for surface and groundwater systems, exacerbated by the increased 'flashiness' of many urban surface water systems,
- densely spaced abstractions which typically change in an unknown manner in time (and sometimes in space), and which can dominate within aquifer flows,
- large changes in system behaviour seen as urban aquifers become heavily exploited,
- heterogeneous land use in space and time resulting in (often unknown) changes in pollutant loadings, which in turn result in heterogeneous distributions of pollutants in space and time throughout the aquifer,
- continuing introduction of a very wide range of chemicals, some with unknown environmental chemistry behaviour,
- pollutants with imperfectly understood transport properties, including NAPLs (non-aqueous phase liquids) and particulates (especially bioparticles), and
- the presence of long-lived, sometimes possibly mobile, dissolved phase pollutant sources (NAPLs, made ground).

These problems can be reduced to three main challenges for modelling:

1. the small space and time scales occupied by many of the processes, e.g.
   a) recharge
   b) contaminant sources
2. the intensity of many of the processes, e.g.
   a) changes in water level (recharge effects, disruption of flow)
   b) degree of contamination
3. lack of knowledge of the physical, chemical and biological processes, e.g.
   a) unstable unsaturated flow
   b) made ground hydraulics and chemistry
   c) leakage hydraulics
   d) (bio)particle and NAPL movement
   e) the chemical behaviour of many pollutants, especially 'new' chemicals.
4. the complexity of code required to deal with surface water systems, pipe network flows, and groundwater.

An *ideal* urban groundwater modelling code, in addition to being quick to run and easy to use, would thus have the following features:

- three-dimensional representation, with the capability to reproduce complex geometrical conditions, including:
   a) connection to GIS input to use land use and other data sources stored in this medium
   b) representation of linear, point and diffuse recharge
- capability to simulate transient flows
- capability to deal with small-scale space and time variation
- unsaturated zone flow representation, with
   a) over-ride for by-pass flows
   b) capability for head-dependent recharge rates
   c) moisture-content-dependent evapotranspiration

d) depth-dependent evapotranspiration
e) variable recharge processes as determined by land use that changes as a function of space and time (possibly via a separate GIS model)
f) capability for perched river conditions resulting from variable river bed sediments
- surface water flow representation to allow comparison with hydrographs
- solute transport capability, with
  a) at least linear retardation and first-order decay for an unlimited number of media, including river bed sediments
  b) concentration boundaries both internally and at the model edges (to represent NAPLs in particular)
  c) ability to track solutes into surface waters
  d) ability to simulate density effects
- representation of wells, including casing depths, to allow flow and solute transport to be assessed for different well designs
- ability to run in stochastic mode.

The importance of these features is largely determined by the purpose of the modelling exercise (and hence the required scale of interest), and by the aquifer characteristics. For example, for regional regulatory investigations, many pollutants can be considered to have diffuse sources, whereas investigations for the development of a new well may require the same pollutants to be considered as having point sources. However, even if a pollutant can be considered diffuse, evaluating its behaviour at the large scale may actually prove at least as challenging as having to consider small-scale transport processes. In general, though, confidence is likely to be greater for predictions at larger scales. As in most other groundwater modelling contexts, stochastic modelling is often an attractive way to attempt to indicate such uncertainties.

When considering the evolution of the groundwater system in a city, or comparing present-day systems in developing cities with those in more established cities, continually advancing technologies and changing social systems mean that the key to the future does not necessarily lie in the past. This is not to say that history is not fundamental to understanding present-day groundwater flow and quality in a given urban aquifer: each urban system is unique – not just because every aquifer has a different physico-chemical composition, but because of the unique history of impacts that heavily concentrated human activity has on the underlying aquifer. It is through taking into account these highly variable and varying impacts that an urban groundwater modelling code can be fundamentally distinguished from other groundwater model codes.

1.4 NUMERICAL MODELLING OF GROUNDWATER IN URBAN AREAS – THE STATE OF THE ART

Groundwater is widely recognized as the planet's largest and most important source of fresh and accessible potable water. It has a vital role to play if the world's rapidly growing cities are to be healthy and sustainable. Where available, groundwater tends to be preferred over surface water in lakes and rivers since it is relatively well protected from pollution, is less prone to drought conditions, and can be introduced incrementally, one pumping well at a time, to meet increasing demand with minimal investment of capital. However, new technologies and well-planned and executed management strategies are critical if groundwater resources are to be used effectively and responsibly. In turn, the development of appropriate management strategies requires a sound knowledge of urban groundwater science, as well as the application of aquifer modelling tools used to test alternative approaches to resource management strategies and aid decision-making (Howard, 2007). Particular consideration needs to be given to the urban water budget and the influence of groundwater capture on induced recharge, and how these, in turn, relate to the sustainable yield of an aquifer (Sophocleous, 2000, 2005, 2007). All these factors are intimately linked and vary quantitatively with time. As such, numerical models represent the only viable approach for evaluating alternative resource management scenarios.

The purpose of this section is to provide a broad background to the *UGROW* model concept by reviewing the current status of groundwater modelling as applied to urban areas. It begins with a very brief history of the numerical approach to aquifer simulation, and continues with a review of existing groundwater models and the extent to which they can cope with the complexities of urban groundwater systems. This review is then used to highlight the important niche that *UGROW* intends to fill.

1.4.1 Developments in numerical modelling

In the past twenty-five years our knowledge of urban groundwater issues and our ability to model them have both advanced significantly. For the most part, developments in these two areas have occurred in parallel. It is only in recent years that these paths have merged, with consideration given to designing models that deal specifically with types of hydrogeological conditions commonly associated with urban areas.

The earliest numerical models were based on the finite-difference method of approximating governing field equations (Rushton and Redshaw, 1979). They were normally two-dimensional, simple in concept, computationally efficient and were typically used to study water resource issues at the catchment scale from a purely quantitative standpoint. Later models began to use the finite-element analytical approach, which offers some advantages over finite-difference techniques, but is mathematically more demanding and consequently more difficult to express in code. Today, finite-difference and finite-element model codes coexist side-by-side. Both take full advantage of fast microprocessors, large memory banks and sophisticated graphical user interfaces (GUIs) to provide highly sophisticated steady-state and transient simulations of groundwater flow and contaminant transport in geologically complex three-dimensional systems under conditions of variable fluid density and changing boundary conditions.

To date, most commercially available groundwater models have been designed in generic form to meet the broader needs of users. For example, commercial environments which can run the USGS finite difference modelling code MODFLOW (McDonald and Harbaugh, 1988, 2003) such as Waterloo Hydrogeologic's 'Visual Modflow' and ESI's (Environmental Systems International's) 'Groundwater Vistas' include:

- MODFLOW/MODFLOW2005 – the most widely used 3D groundwater flow models in the world, capable of representing the effects of wells, rivers, streams, drains, horizontal flow barriers, evapotranspiration and recharge on flow

- MT3D – a 3D contaminant transport model that can simulate advection, dispersion, sink/source mixing and chemical reactions of dissolved constituents
- MODPATH – a 3D particle-tracking model that computes the path a particle takes in a steady-state or transient flow field over a given period of time
- PEST – parameter estimation and automatic calibration.

These models also include support for more advanced developments:

- MODFLOW-SURFACT – incorporates flow in the unsaturated zone, delayed yield and vertical flow components
- SEAWAT – can simulate three-dimensional, variable-density, transient groundwater flow in porous media.

These features considerably improve our ability to model the flow of groundwater and entrained contaminants in urban areas. However, none are explicitly designed to deal with certain specific and peculiarly problematic characteristics of the urban subsurface environment, such as leaky pipes and utility trenches, storm drains, poorly compacted fill and large underground openings, for example, tunnels and excavations, that collectively contribute to what has been described as 'urban karst' (Sharp et al., 2001; Krothe, 2002; Krothe et al., 2002; Sharp et al., 2003; Garcia-Fresca, 2007).

In a similar vein, FEFLOW®, a popular and highly versatile finite-element modelling code developed by WASY GmbH, Berlin, is also capable of simulating flow and transport processes in three-dimensions and can handle:

- dual porosity media,
- saturated and unsaturated conditions,
- mass and/or heat transport,
- chemical reactions and degradation mechanisms,
- variable fluid density problems resulting from variable salt concentration and/or influences of temperature, and
- time variant boundary conditions.

However, although these characteristics are frequently encountered in urban aquifer systems, the onus is once again placed on the model user to adapt the model to deal with the often unique complexities of the urban subsurface.

1.4.2 The interim solution

Until recently the solution to the problem of modelling urban groundwater systems has been to either:

1) use available groundwater flow and transport models and adapt the urban data in such a way as to meet the modelling requirements, i.e. the 'adaptation' approach, or
2) develop independent models of the urban water system or of the urban water system components and 'marry' these urban system models to existing flow and transport models, i.e. the 'coupling' approach.

Examples of these approaches are provided below.

The adaptation approach

As an example of the adaptation approach, Visual Modflow has recently been used to investigate the potential impacts of urban growth in the Greater Toronto Area, southern Ontario, Canada (Howard and Maier, 2007). The study area – the Central Pickering Development Lands, better known as the Seaton Lands – is part of the Duffins Creek Drainage Basin, a catchment comprehensively studied and modelled with MODFLOW by Gerber and Howard (1996, 2000, 2002) in an effort to understand the hydrogeological behaviour of the Newmarket Till aquitard. The model developed as part of this work was subsequently adopted with some modification for use in the impact analysis. It includes nine layers with a grid discretization of 200 m by 200 m (110 columns and 150 rows), and was configured using borehole data from approximately 7,000 Ontario Ministry of the Environment water well records, supplemented by borehole data from landfill and regional water resource investigations. Full details of the model and the steady-state calibration are provided by Gerber and Howard (2002).

The Seaton Lands are shown in Figure 1.6. These are a primary target for the next phase of major urban development in the Greater Toronto Area (GTA), and to many this

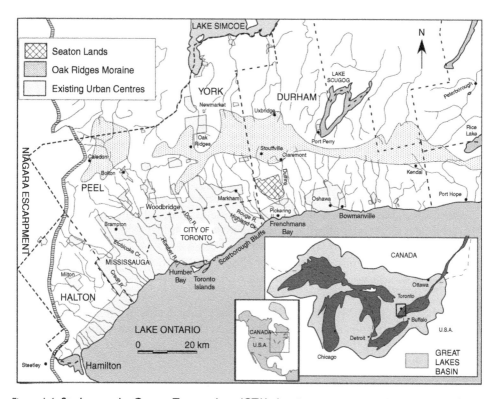

Figure 1.6 **Study area: the Greater Toronto Area (GTA) showing component regions, existing urban centres and the Seaton Lands**

Source: After Howard and Maier, 2007

development is seen as an important test of the provincial government's resolve to ensure that growth in the GTA is environmentally sustainable. Primary issues regarding these lands include the effects of development on the local water balance, the fate of sensitive wetlands, and the potential impacts of NaCl road de-icing salts on groundwater quality.

In adapting Visual Modflow to investigate the potential impacts of road de-icing practices on local aquifers, it was necessary to make two important assumptions to overcome limitations in the basic model formulation:

- Salt released to the subsurface is immediately transferred to the water table. In other words, the potential role of the unsaturated zone not simulated in MODFLOW can be ignored. This assumption was considered acceptable given that vertical travel times for salt to the shallow water table are likely to be small when compared to travel times in the aquifer.
- Major arterial roads and highways that would normally behave as line sources of salt contamination can be represented as multiple point sources of contamination, the salt being introduced as mean annual concentrations at the nearest model node. This assumption was considered to be acceptable given the regional scale of the study.

For predictive purposes, it was also assumed that current rates of salt application will be maintained in the future.

The model was then developed in two stages:

- The model was run to steady state in the absence of salt loadings to recreate the previous modelling work performed by Gerber (1999).
- Salt was applied to existing roads and highways in the watershed for a period of fifty years to represent the historical loading of salt on the system. This effectively represents the pre-development condition for the Seaton Lands.

Subsequently, salt impact predictions were obtained for two scenarios:

- The long-term, chemical steady state in the absence of any development, in other words, the impact that will accrue when the mass of salt entering the system in recharge will be in equilibrium with the mass of salt lost as discharge to streams, rivers and Lake Ontario (Figure 1.7).
- The long-term, chemical steady state should the Seaton Lands be developed and thus impose additional loadings of salt to the system (Figure 1.8).

Despite the inability of MODFLOW to represent finer details of the urban system at the local scale, the results are considered useful at the regional scale, and do provide a valuable indication of the potential impacts of development and the time-frame over which they will occur.

The coupling approach – OROP

One of the first successful, applied examples of the coupling approach was the Optimized Regional Operations Plan (OROP) (Figure 1.9), pioneered and implemented in 1999 by Tampa Bay Water, Florida's largest wholesale water supplier (Hosseinipour,

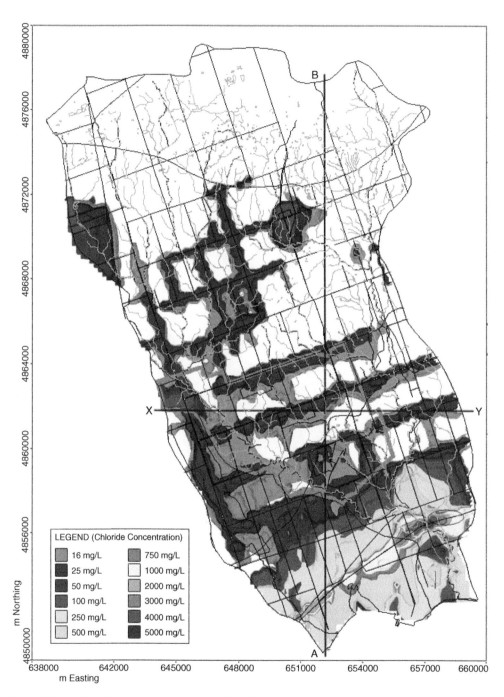

Figure 1.7 Predicted long-term, steady-state chloride concentrations in the uppermost aquifer in the absence of urban development in the Seaton Lands study area. Note that while chemical steady state is not achieved for several hundred years, most of the change occurs within a time-frame of about 100 years (see also colour plate 3)

Source: The authors

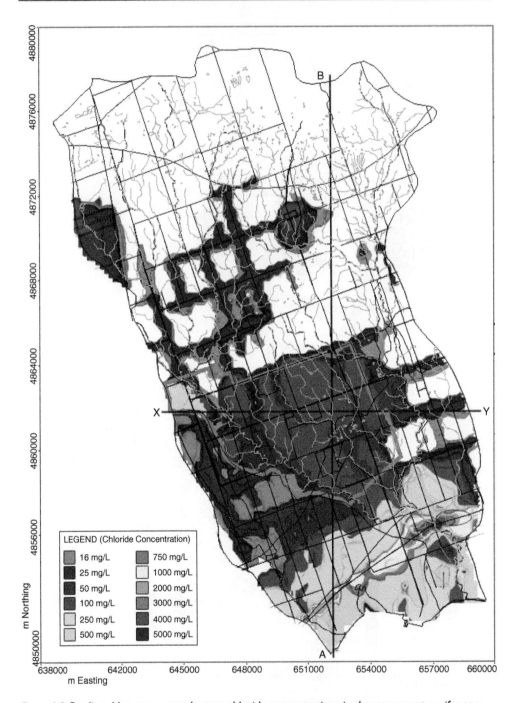

Figure 1.8 Predicted long-term, steady-state chloride concentrations in the uppermost aquifer as a result of road salt application following development of the Seaton Lands study area. Note that while chemical steady state is not achieved for several hundred years, most of the change occurs within a time-frame of about 100 years (see also colour plate 4)

Source: The authors

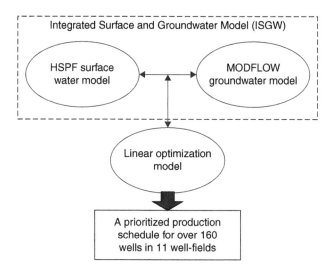

Figure 1.9 The Optimized Regional Operations Plan (OROP)

Source: The authors

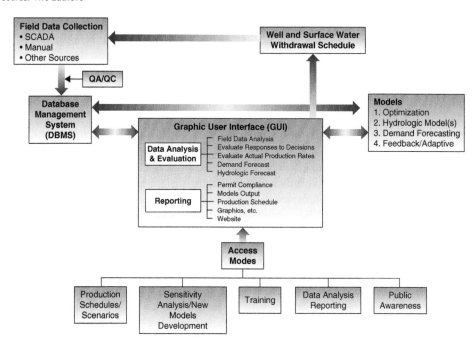

Figure 1.10 Components of the decision support system as envisioned by Hosseinipour

Source: Hosseinipour, 2002

2002). Tampa Bay Water's member governments include the cities of New Port Richey, St. Petersburg and Tampa, and the OROP is a key component of the local water resource Decision Support System (DSS) (Figure 1.10), serving more than 2 million residents in the Tampa Bay region.

The OROP is a customized computer tool that uses forecasted surface water availability, rainfall data and current water-level conditions and operating constraints to determine how to rotate production among available well-field supplies to meet demands in an environmentally sound manner, for example, preventing wetland deterioration and sea water intrusion. It combines an integrated surface–groundwater simulation model with an optimization program to produce a prioritized production schedule for over 160 wells in eleven well-fields.

A key component of OROP is the integrated hydrologic simulation model of the Central Northern Tampa Bay area (CNTB) developed by SDI Environmental Services, Inc. (SDI). The CNTB model employs SDI's ISGW (Integrated Surface water Ground Water) software, which couples surface water hydrology, as simulated by the HSPF model (Hydrologic Simulation Program) (Johanson et al., 1984), and the aquifer system as simulated by MODFLOW (McDonald and Harbaugh, 1988). Given OROP's vital role in Tampa Bay Water's well-field management system, understanding and quantifying the uncertainty of predictions made by the CNTB model is recognized as a high priority for ensuring long-term sustainability of the water resource.

MODFLOW is used to simulate groundwater flow in two principal hydrostratigraphic units: the generally unconfined Surficial Aquifer System (SAS) and the Upper Floridan Aquifer System (UFAS), which is confined over much of the model area by the Intermediate Confining Unit (ICU). The model consists of two layers (the SAS and the UFAS), 153 rows and 152 columns with cell dimensions ranging from one-quarter mile (about 400 m) to 1 mile (about 1,600 m). The Intermediate Confining Unit (ICU) is not explicitly discretized within the MODFLOW model and hundreds of ICU 'windows' are probably present over the model domain. These holes are normally formed by collapses or sinkholes and will usually result in a direct hydraulic connection between surface water bodies and the UFAS hydrologic systems.

The modelling approach adopted for OROP is clearly very limited in its ability to represent important characteristics of urbanized areas such as urban 'karst' and multiple, often very discrete, sources of aquifer recharge. OROP succeeds because of the large scale of the model and the very broad resource issues the model is asked to address. This type of model would clearly be inappropriate for investigating the urban water balance and urban water quality impacts at a very local scale. Such tasks would be more suited to the AISUWRS model, as described below.

The coupling approach – AISUWRS

One of the most advanced examples of the 'coupling approach' is the AISUWRS model (Assessing and Improving Sustainability of Urban Water Resources and Systems). This modelling tool involves a series of intimately linked urban component models. It was developed during a three-year international multidisciplinary research project, funded by the European Commission under the Fifth Framework Programme, the Department of Education, Science and Training of Australia and the UK Natural Environment Research Council. Participating organizations included:

- University of Karlsruhe, Germany (Coordinator),
- British Geological Survey, United Kingdom,
- Commonwealth Scientific and Industrial Research Organisation, Australia,

- FUTUREtec Gmbh, Germany,
- GKW Consult, Germany,
- Institute for Mining & Geology (IRGO), Slovenia, and
- University of Surrey (Robens Centre for Public and Environmental Health), United Kingdom.

Full details of the model are published in the comprehensive 'Urban Water Resources Toolbox' (Wolf et al., 2006b).

The AISUWRS project was driven by the need to integrate groundwater into urban water management. At the heart of the project was the recognition that while land and water use in urban areas is highly complex, cities are expanding and municipal water utilities can no longer afford to neglect water in an underlying productive aquifer just because it is difficult to assess.

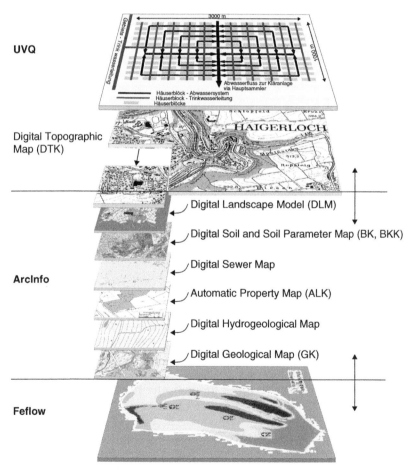

Figure 1.11 AISUWRS model concept as conceived by Eiswirth

Source: Eiswirth, 2002

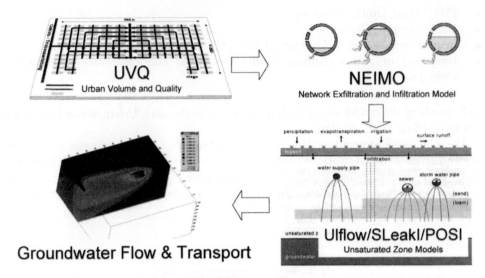

Figure 1.12 Major model compartments demonstrating the integrated approach of AISUWRS (see also colour plate 5)

Source: After Wolf et al., 2006b

As originally conceived (Eiswirth, 2002), the AISUWRS model linked an existing Urban Volume and Quality model (UVQ model), developed through Australia's CSIRO urban water programme (UWP) with a groundwater flow model (FEFLOW®) via a series of (ARCINFO®) GIS layers (Figure 1.11). Since then, the AISUWRS project has developed a range of other component modules including the Network Exfiltration and Infiltration Model (NEIMO) and a series of unsaturated zone models that now provide the interface between the UVQ and the model (Figure 1.12). The fully linked system including the decision support system (DSS) and the Microsoft Access database is shown in Figure 1.13.

1.4.3 The niche for *UGROW*

Despite its demonstrated ability, the AISUWRS modelling tool has enjoyed only limited success. This is explained in part by its high level of sophistication and high data demands, which limit its application to cities that either have a well-established database or sufficient funds to acquire the necessary data. A second, more fundamental problem relates to the 'coupling approach' adopted by AISUWRS. This places ultimate reliance on an independently developed groundwater flow model, such as the finite-element model FEFLOW, to complete the package and deliver the results. In practice, problems have been encountered in obtaining a seamless 'marriage' of AISUWRS model components with FEFLOW (Wolf et al., 2006b), and only time will tell if such issues can be fully resolved.

As a fully integrated model dedicated to urban water systems, *UGROW* exploits the perceived failures in AISUWRS by providing a complete and fully integrated modelling package for simulating urban water systems. Certain components of *UGROW* may

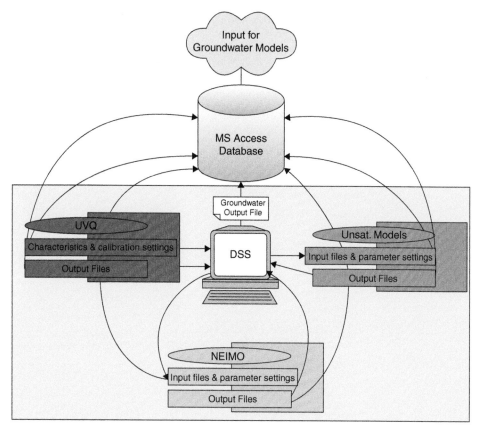

Figure 1.13 Link between the key AISUWRS model components, the decision support system (DSS) and the Microsoft Access database (see also colour plate 6)

Source: After Wolf et al., 2006b

lack some of the flexibility that AISUWRS offers, but all key urban water system elements are represented and seamlessly linked. The current version of *UGROW* has some limitations, not least its inability to simulate complex multi-layer aquifer systems. Nevertheless, it represents a viable, user-friendly urban water simulation package that is readily applicable to a wide range of urban water resource issues. As such, the model fills a valuable niche in the tools available to urban water system managers and decision-makers.

Chapter 2

UGROW – the Urban GROundWater modelling system

Dubravka Pokrajac[1] and Miloš Stanić[2]

[1]School of Engineering, University of Aberdeen, Aberdeen, United Kingdom
[2]Institute of Hydraulic Engineering, Faculty of Civil Engineering, Belgrade, Serbia

The main concepts and rationale that motivated the development of the *UGROW* software are presented in Chapter 1. In Chapter 2, we present an overview of the *UGROW* model system and its key attributes. Section 2.1 defines the scope and limitations of the system's application, detailing the types of practical problems that can and cannot be solved using *UGROW*. Section 2.2 explains the theoretical basis of the model, including details of its components. Section 2.3 describes the groundwater simulation model *GROW*, while Sections 2.4 and 2.5 contain details of the unsaturated flow model *UNSAT* and the surface runoff model *RUNOFF*, respectively. Section 2.6 lists the data required to describe a real-world problem and run *UGROW*; and Section 2.7 focuses on the Graphical User Interface (GUI). It explains the computational geometry algorithms used for handling geographical data and presents a simple hypothetical case study providing a step-by-step procedure for model development. Finally, Section 2.8 briefly outlines model calibration, uncertainty and sensitivity; these elements are important for the practical application of any groundwater simulation model and are not specific to *UGROW*.

2.1 MODEL CONCEPTS

2.1.1 General features

The management of urban groundwater is an integral component of a well-managed urban water system. Chapter 1 describes the key features and special needs of integrated management of urban water systems and urban groundwater in particular. In this section we introduce *UGROW*, an urban water management tool designed to address many of these needs.

UGROW is a software tool dedicated to the management of the Urban GROundWater component of urban water systems. The software system was developed to raise awareness of the interaction between urban groundwater and other urban water systems, and to improve the capability of simulation models to represent this behaviour. Its main purpose is to allow the interaction of urban groundwater with other urban water systems to be visualized, demonstrated and quantified. To fulfil this task, vast amounts of data describing various urban water systems need to be stored and efficiently manipulated. This became feasible relatively recently with the rapid advent of powerful desktop computing resources that permit the design and development of a new generation of simulation models that can perform highly complex tasks. In *UGROW*, sophisticated, dynamic, simulation models are linked with GIS (Geographical Information Systems) to provide one of the most advanced urban groundwater simulation systems currently available.

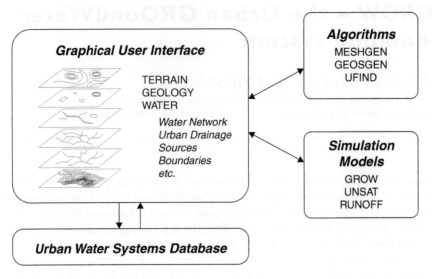

Figure 2.1 Basic structure of UGROW

Source: The authors

The basic structure of *UGROW* is shown in Figure 2.1. The main components of *UGROW* are a graphical user interface for data manipulation, a GIS urban water systems database, a set of algorithms for data manipulation and three simulation models.

2.1.2 User interface

The user interface of *UGROW* is called *3DNet*, because it is dedicated to the **3D** (three-dimensional) presentation of various urban water systems, mostly in the form of **Networks**. It is a Microsoft Windows-based graphical pre/post-processor with Geographical Information System (GIS) functionality. This means that the information is organized within a series of layers, which can be superimposed and graphically presented (Figure 2.2). However, unlike conventional GIS packages, graphical objects such as contours of terrain, pipes, streams and model boundaries, and so on, are truly three-dimensional. The *3DNet* user interface communicates with the database by reading graphical objects from, or writing them into, the database. It also manipulates the three-dimensional view and the plan view of a drawing scene as shown in Figure 2.2.

The *3DNet* user interface is an integrated hydro-informatics tool which contains three key components:

- *TERRAIN* is dedicated to the presentation and handling of the ground surface. It:
  - inserts and fits scanned maps,
  - inserts digitized or imported elevation points and terrain contours,
  - creates a Digital Terrain Model (DTM) by Delaunay triangulation, and
  - creates contour lines.
  It uses predefined or customized colour maps for DTM presentation.

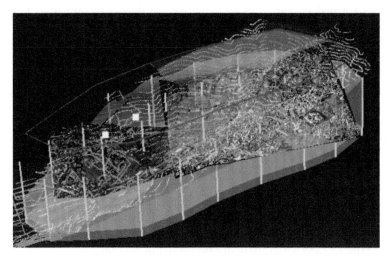

Figure 2.2 A 3D view of a terrain model and the hydrogeological layers (see also colour plate 7)
Source: The authors

- *GEOLOGY* handles the geological layers defined by a series of boreholes, with each borehole containing several layers. The boreholes can be:
  - 'real', using recorded, site-specific borehole logs, or
  - 'fictive', inserted to produce a realistic representation of interpreted geological layers.

  The *GEOLOGY* component manipulates real and fictive borehole data and implements an algorithm for creating geological layers.

- *WATER* is used for operating all water systems and water simulation models. It creates the water-supply network (*WATNET*), the urban drainage network (*SEWNET*) and an urban streams network (*STREAMNET*). It also manipulates all data needed to run the available simulation models, which include:
  - *RUNOFF* for the balance of surface runoff,
  - *UNSAT* for simulation of unsaturated flow, and
  - *GROW* for groundwater flow simulation.

  For the purpose of groundwater simulation, the *WATER* component defines the hydrogeological units (the primary aquifer and, where present, an overlying aquitard) and the model domain boundaries; it generates the finite-element mesh using the algorithm *MESHGEN* and connects urban water networks to the groundwater simulation model using the algorithm *UFIND*.

2.1.3 The database

The urban water systems database stores data on the terrain, geological layers, water systems and groundwater model for a particular city or part of a city. Terrain data include a series of (x, y, z) coordinates of terrain points, lines connecting the points and the triangles formed by the lines to generate the digital terrain model. Geological layers are created between digital terrain models of their upper and lower surfaces as defined by the (x, y, z) coordinates of points extracted from borehole data. The digital models for the geological layer surfaces are formed from these points using the same algorithm as used for the digital terrain models. The space between the models of these surfaces is then filled by a solid-generating algorithm called *GEOSGEN*.

Water systems that can be stored in the database include: water supply pipes, sewers, streams and wells. Each system comprises a series of objects (e.g. pipes) and each object has several *attributes,* for example, pipe length, diameter, operating water level (for sewers) and pressure head (for water supply mains). All objects can be drawn on the computer screen and for that purpose have *properties* (e.g. line colour, thickness, text size and text colour).

The database also contains data for the simulation models, such as time-series for effective precipitation, runoff coefficients, physical boundaries and boundary conditions, and so on. The general data requirements and the detailed list of data that can be stored in the *UGROW* database are presented in Section 2.6.

2.1.4 Algorithms

Algorithms describe the procedures used to handle the data and pass the correct information to the simulation models. *UGROW* contains a library of algorithms used to integrate its many components. The following algorithms are available:

- *MESHGEN* performs mesh generation within a given domain. Using Delaunay triangulation, *MESHGEN* divides the domain into triangles which cover the domain in its entirety without overlap. A sample mesh generated using *MESHGEN* is shown in Figure 2.3. The density of this mesh is controlled by specifying a maximum limit to the area of a triangle. The domain boundary consists of either predefined segments, or predefined points and lines. The former method is used for generating a finite-element mesh within the groundwater flow modelling domain. Points and lines are first enclosed within a convex hull (the smallest convex surface that contains all of them) and the mesh is generated within this hull. The latter method is used to generate the digital terrain model.
- *GEOSGEN* performs the solid generation required to represent geological layers and hydrogeological units (Figure 2.4). The geometry of a solid is defined by subdividing the space between the two predefined surfaces into tetrahedrons using Delaunay tessellation. This procedure is a three-dimensional analogy to triangulation in a plane. An example is shown in Figure 2.4.
- *UFIND* searches the database for water system objects that are potential sources of groundwater recharge, and assigns them to the individual elements of the finite-element mesh of the groundwater simulation model. This is a crucial algorithm for integrating all the water systems contained in the *UGROW* database.

2.1.5 Simulation models

The simulation models provide a numerical representation of the movement of water for key parts of the hydrological cycle: surface runoff, subsurface flow in the unsaturated

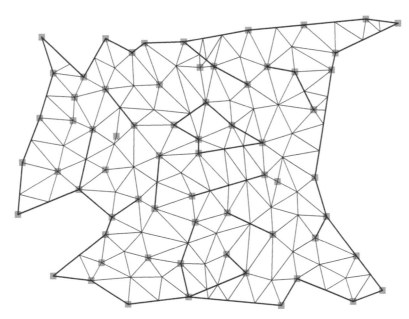

Figure 2.3 Mesh generated by MESHGEN

Source: The authors

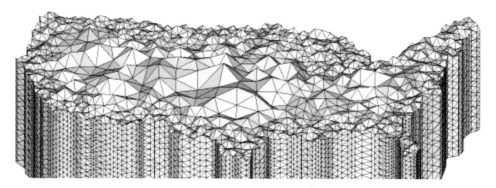

Figure 2.4 A solid model of a geological layer generated using GEOSGEN

Source: The authors

zone and groundwater flow. The behaviour of other water system components is defined through their operational parameters, but is not dynamically simulated.

The key simulation models can be considered modules of *UGROW* and include:

- *GROW,*
- *UNSAT,* and
- *RUNOFF.*

GROW is a groundwater flow model that simulates flow and contaminant transport in an urban aquifer recharged from natural sources, such as precipitation, and from artificial sources, such as leakage from water mains, stormwater sewers and landfills. Part of the aquifer may be unconfined and directly connected hydraulically to the ground surface, while other parts may be isolated from the surface by an intervening aquitard. In either case, the uppermost part of the soil near the ground surface is unsaturated, and infiltration either seeps through this unsaturated zone and recharges the aquifer, or is extracted by capillary forces and returned to the atmosphere as evapotranspiration. In urban areas, aquifers also receive water via leakage from water supply pipes, sewers and so on, and release water as baseflow to streams and discharging abstraction wells. The groundwater simulation model uses the Galerkin finite-element method to solve the basic groundwater flow equation. The equation contains source and sink terms which account for the interaction of groundwater with the urban supply and sewer network. Results of a simulation include a time-series of groundwater levels (heads) and individual components of the groundwater balance such as time-varying recharge rates from the unsaturated zone, water supply pipes and sewers.

The unsaturated flow module, *UNSAT*, replicates vertical soil moisture flow immediately below the ground surface. Soil parameters are selected according to the land surface and land use. The result of the simulation is a time-series of soil moisture profile and vertical downwards flux at the base of the predefined unsaturated zone or, for a shallow unconfined aquifer, at the water table. This flux is used by the groundwater simulation model as the aquifer recharge.

RUNOFF, the runoff simulation model, automatically divides the ground surface into catchment areas and tracks the surface runoff, calculated from estimates of effective rainfall and the runoff coefficient, to the receiving sewer or stream. The runoff simulation model generates a water balance for any given point along a stream or a sewer network.

2.1.6 Using UGROW

Working with *UGROW*, a user can store and visually inspect all data for the different water systems. The three-dimensional graphical post-processor becomes a 'virtual reality' tool that can depict the complexities of the urban subsurface. An example of such a visualization is shown in Figure 2.5.

After visual inspection of the data, the simulation models are run to calculate and predict the movement of water in the urban environment. Throughout the simulation, groundwater interaction with other systems is continuously updated. The results of the simulation are viewed using the graphical post-processor. By superimposing groundwater levels on the pipe network, it is very easy to identify potential problems, for example, risk of groundwater contamination by leaking sewers. Graphical presentation of groundwater levels integrated with the water systems is also valuable for educational purposes, since it clearly demonstrates how these systems interact.

2.2 MODEL APPLICATION

2.2.1 Physical model

Figure 2.6 shows a typical physical system that can be simulated using *UGROW*. It comprises a land surface covering either part or all of a city, subsurface hydrogeological units and urban water network features such as water supply mains, sewers, wells,

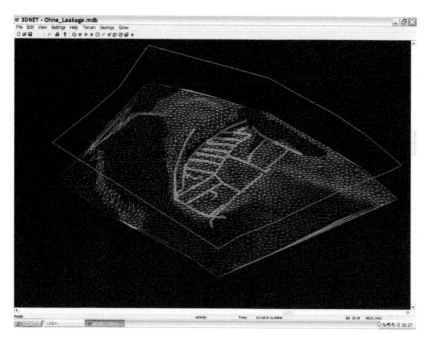

Figure 2.5 View of a water supply pipe and sewer in the city of Rastatt. The details of the case study are presented in Section 3.1 (see also colour plate 8)

Source: The authors

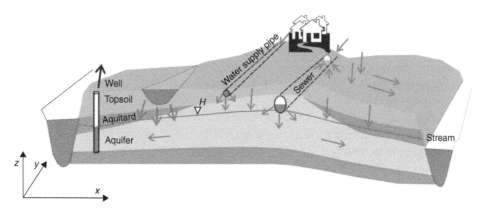

Figure 2.6 A typical physical system that can be simulated using *UGROW* consists of a land surface with various land uses, an aquifer, upper and lower aquitards, an unsaturated zone, water supply mains, sewers, wells, streams and other urban water features (see also colour plate 9)

Source: The authors

streams and so on. The main part of the system is the urban aquifer, which is one of four hydrogeological units. Other units are the unsaturated zone within the soil near the land surface (referred to here as 'topsoil'), and two aquitards, one overlying the aquifer as a confining layer, and the other underlying the aquifer and forming a base. All hydrogeological units are layers, that is, their thicknesses are much smaller than their horizontal dimensions; however, they are generally variable within space and can be locally discontinuous. For example, the overlying aquitard may confine and protect the aquifer in some areas but be absent in others. Additional urban water components include objects (pipes, stream sections, wells etc.) which are mostly buried within the uppermost metres of the urban subsurface. Some of these objects may be literally inside the aquifer, while others may have indirect contact with it.

It is well established (Chapter 1) that altering land use and the operation of urban water systems can have a major effect on the natural hydrological cycle. In urban areas, land use is highly variable. Vegetated areas are permeable to recharge while paved areas are normally impervious and create significant runoff. Runoff follows the topography and feeds urban sewers and surface water features such as streams. Water that infiltrates into the permeable soil passes through the unsaturated zone to recharge the aquifer. Urban sewers, water supply mains, streams and similar systems may provide additional recharge or, in some cases, act as drains. Even a single pipe can behave as a source of recharge in one urban region and a sink or a drain in another. Similarly, a single pipe may be a source of recharge during the dry season with low groundwater levels and become a drain during the wet season when high groundwater levels prevail.

UGROW can simulate unsteady-state flow in urban aquifers that strongly interact with other urban water systems. A typical simulation result contains groundwater levels and water balance terms, including groundwater recharge/discharge from the vadose zone and inflow/outflow from sewers, water supply mains and similar urban water system components.

2.2.2 The urban water balance

The ability to represent the transient behaviour of individual components of the urban water balance is an important feature of *UGROW*. The hydrological cycle in urban environments and key water balance components were briefly introduced in the previous section. Here, they are examined in detail. Within the urban environment, we can consider a series of 'control volumes' as follows:

- land surface,
- soil zone (unsaturated zone or 'topsoil'),
- water supply network,
- sewer network,
- stream network,
- ponds and landfills,
- wells,
- other point, line or areal sources/sinks of water, and
- aquifer.

Any of the control volumes can be divided into a series of sub-volumes; for instance, a pipe network can be separated into individual pipes and the groundwater simulation model domain can be broken into individual elements. As a general principle, the balance equation:

Inflow – Outflow = Storage   (2.2.1)

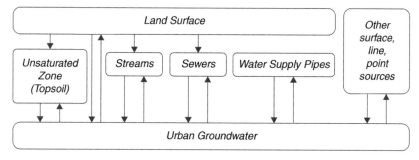

Figure 2.7 The interaction between control volumes in the urban water balance

Source: The authors

can be applied to any control volume (or sub-volume) for an arbitrary time interval Δt, where *Inflow* is the volume of water (or volume of water per unit time $\times$ Δt) that enters the control volume, *Outflow* is the volume that leaves the control volume and *storage* is the change in the volume stored. The physical meaning of the terms in the balance equation depends on the actual control volume in question. For example, within the soil zone (herein often referred to as the unsaturated zone or 'topsoil'), *Inflow* refers to the part of the rainfall which infiltrates the soil, *Outflow* is the volume of water that leaves the soil to enter the aquifer as recharge, and *Storage* is the volume of water that accumulates in the soil and is reflected by a change in its water content.

The interaction between different control volumes in the urban water balance is schematically represented in Figure 2.7. Although all components of the water system potentially interact, quantifying this interaction in a real case scenario can be extremely difficult, partly because most water systems are buried in the ground and hence not accessible, and partly because some of the parameters that influence the interaction are difficult to evaluate.

Since the focus of *UGROW* is urban groundwater, detailed dynamic simulation is performed only for the aquifer and the overlying unsaturated soil. Other water systems are included with only enough detail to represent their influence on groundwater. In other words, water balance calculations for such systems are performed at the scale appropriate for the groundwater balance but not with the level of detail needed to investigate their operation. The remaining part of this section focuses on the urban groundwater balance.

A series of simulation models or 'modules' are built into *UGROW* to represent components of the water balance. These, as indicated in Section 2.1, include:

- *GROW* for groundwater flow and its interaction with various urban water system components such as water supply pipes, sewers and wells,
- *UNSAT* for water seepage through the soil/unsaturated zone close to the ground surface, and
- *RUNOFF* for tracing overland flow.

Figure 2.8 illustrates the role of each model within the urban water balance simulation.

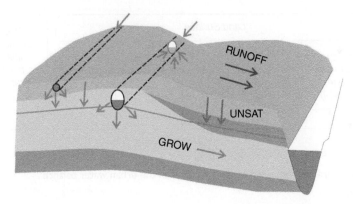

Figure 2.8 The three simulation models related to physical processes in the urban water balance (see also colour plate 10)

Source: The authors

Throughout this book, the sign conventions used to indicate groundwater balance are as follows: the '+' sign always indicates groundwater recharge, that is, inflow into the aquifer, and the '−' sign indicates groundwater drainage, that is, outflow. The interaction between groundwater and any other source is always termed 'groundwater recharge' with the understanding that in an actual simulation, a negative result for this term will mean that the 'source' is behaving as a sink and is draining groundwater. The same convention is used for prescribing all model boundary conditions, for example, the discharge from an abstraction well is represented as a negative value.

At the surface, the portioning of precipitation between surface runoff and potential soil infiltration is defined by:

$$P_{eff} = P(1 - C_{sr}) \tag{2.2.2}$$

where P is net precipitation (precipitation − interception by plants), C_{sr} is the surface runoff coefficient and P_{eff} is effective precipitation, which has the potential to infiltrate the soil either during or after rainfall as a function of soil permeability. The subsequent migration of infiltrated water through the soil is calculated using the unsaturated flow simulation model *UNSAT*. One of the outputs from this calculation is groundwater recharge from the soil into the underlying aquifer. Details of the theoretical background underlying the *UNSAT* model, as well as the numerical model features, are presented in Section 2.4.

Surface runoff flows across the land surface until it reaches a recipient – a sewer or an open channel or stream. Its behaviour is modelled using a simplified approach that avoids the dynamic detail of water flow over the land surface, and simply traces the runoff along the surfaces of the digital terrain model with relatively low precision that is still sufficient to capture the key characteristics of the urban water balance. The details of the algorithm that performs the tracing are presented in Section 2.5.

Groundwater flow, including the interaction of groundwater with each and any of the water supply system components is determined within the groundwater simulation model *GROW* (Figure 2.8). The underlying theory and details of the numerical model incorporated in the *GROW* model are presented in Section 2.3. Components or 'objects' of an urban water system can be classified as either surface (e.g. pond), linear (e.g. sewer) or point (e.g. septic tank) sources of groundwater recharge. All these potential sources of groundwater recharge are termed 'Leakage Sources'. Groundwater recharge from any of the Leakage Sources may or may not depend on the aquifer's groundwater level (head) and the hydraulic head at the source. In the former case, a Leakage Source is called 'head-dependent', while in the latter case it is called 'head-independent'. Some sources may be both. For example, if detailed data on the state of individual 'head dependent' sewers are not available, we may assign single values for groundwater recharge from various parts of the sewer network, depending on their age and anticipated level of decay. In effect, there are two general options for evaluating groundwater recharge from water system objects:

- known recharge (volume per unit time) per unit area (for surface source), per unit length (for linear source), or per unit object (for point sources), and
- known relationship between recharge, groundwater level and the elevation of the object and its hydraulic head.

These relationships are discussed in detail in Section 2.3.3.

The groundwater model performs the main simulation, calculates all components of the urban groundwater balance and stores this information in the model output data.

2.2.3 Scope of application

UGROW is dedicated to solving practical problems related to urban groundwater. Typical practical examples include:

- evaluating the risk of groundwater pollution due to leaking sewers,
- evaluating the risk of contaminating water supply mains by polluted groundwater,
- identifying the hydrological conditions under which groundwater may enter parts of the sewer network and thereby increase flows at the treatment plant,
- optimizing strategies for solving the problem of rising groundwater levels due to recharge from leaking water mains,
- optimizing the number, location and operating schedules of abstraction wells, and
- preparing demonstration cases for raising awareness among water authorities, local authorities and practising engineers of potential urban groundwater problems.

The scope of *UGROW* is limited to cases which correspond to the physical system described in Section 2.2.1. Although *UGROW* contains data on a range of water systems (pipes, rivers etc.), *UGROW* does not dynamically simulate their operation. In its current form, the groundwater simulation model is limited to cases that embrace a single aquifer with:

- a rigid porous matrix,
- large horizontal dimensions compared to the aquifer thickness, such that vertical components of flow are negligible compared to horizontal components, and
- principal axes of anisotropy in the horizontal plane that coincide with the coordinate axes.

Examples of problems where it would be inappropriate to use *UGROW* include:

- multiple aquifers, unless they can be replaced with a single equivalent aquifer, and
- aquifers with fractured porosity, unless the scale of the model allows the fractured aquifer to be replaced with an equivalent aquifer with inter-granular porosity.

2.3 *GROW*: GROUNDWATER FLOW SIMULATION MODEL

2.3.1 Introduction

The groundwater simulation module, *GROW*, simulates transient flow in urban aquifers. The physical model of such a system is described in Section 2.2.1 (Figure 2.6). It consists of a single aquifer overlying an aquitard or an impermeable base. The aquifer may be partly or fully covered by an aquitard. It may receive recharge from various sources such as sewers, water supply mains and infiltration wells, and supply water to abstraction wells or drains. It may also be connected to urban streams with recharge or discharge sections.

The two key types of hydrogeological profile simulated by *GROW* are shown in Figure 2.9. These are:

- a two-layered porous medium comprising an aquifer unit and a confining layer, thickness l_{top} and relatively low hydraulic conductivity K_{top}, overlying the aquifer unit (Figure 2.9a), and
- a one-layered porous medium, comprising an aquifer unit with no confining layer (Figure 2.9b).

In either case, the layer underlying the aquifer may either be impervious or consist of an aquitard, simulated as a source of recharge with either known recharge rates or a recharge rate determined by a potentiometric head and given coefficients of hydraulic resistance.

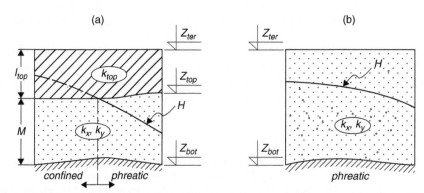

Figure 2.9 Types of aquifer simulated by *GROW*

Source: The authors

Figure 2.9 also shows the notation used for the parameters required to define the geometry of the hydrogeological units: Z_{ter} is the terrain level (land elevation), Z_{top} is the elevation of the top of the aquifer unit, Z_{bot} is the elevation of the base of the aquifer unit, $M = Z_{top} - Z_{bot}$ is the thickness of the aquifer unit and l_{top} is the thickness of the overlying aquitard.

Depending on the hydraulic head H, relative to the top of the aquifer unit Z_{top}, the type of flow in the aquifer can be:

- confined, if the hydraulic head is above the top of the aquifer unit: $H > Z_{top}$ (left-hand side of Figure 2.9a). In this case, the saturated thickness of the aquifer B is equal to the full thickness of the aquifer unit $B = Z_{top} - Z_{bot} = M$, and the hydraulic head is referred to as the potentiometric head, or
- phreatic, if the hydraulic head is lower than the top of the aquifer unit: $H < Z_{top}$ (right-hand side of Figure 2.9a), or there is no confining layer at all (Figure 2.9b). The saturated aquifer thickness is $B = H - Z_{bot}$, and the hydraulic head is generally referred to as the groundwater table or simply 'water table'.

Where aquifer units are overlain by an aquitard, groundwater flow can be confined in one part of the flow domain and phreatic in the other, as shown in Figure 2.9a.

In *GROW*, groundwater flow in the aquifer is described by a two-dimensional mathematical model, based on the fundamental equations of groundwater flow. As a starting point, the mass balance and momentum balance equations are averaged over a representative volume of the porous medium, to form a three-dimensional mathematical model. Further averaging over the saturated aquifer thickness B leads to the two-dimensional model. Averaging over B also requires that boundary conditions at the top and bottom of the aquifer be defined including recharge from overlying aquitards, underlying aquitards and from any other external sources of recharge.

2.3.2 Basic equations

Conventional groundwater flow equations describe water seepage through porous materials which contain just two phases, water and the solid phase (soil grains). The equations are derived by applying spatial averaging to fundamental equations governing the movement of water at the microscopic level (scale of a fluid particle). The averaging is performed over a representative volume that is large enough to ensure that the result does not depend on the size of the volume, and is small enough to exclude the effect of large-scale soil heterogeneities. The volume that satisfies these requirements is the Representative Elementary Volume (REV) (Figure 2.10). The result of the averaging is assigned to the centre of the volume, denoted by x_0 in Figure 2.10. By sweeping the domain of interest with a REV of the scale of engineering interest (e.g. the scale of a soil sample), microscopic equations of water movement are replaced with macroscopic equations which now contain macroscopic parameters. While at the microscopic level equations are defined only within volumes occupied with water (e.g. at the point x in Figure 2.10), macroscopic equations are valid everywhere. A thorough explanation of the averaging procedures, as well as the detailed derivations of the basic equations presented in Sections 2.3.2 and 2.4.1 are presented in Bear and Bachmat (1991). For simplicity, the spatially averaged (i.e. macroscopic) variables are distinguished from the microscopic variables only descriptively, and the special notation for spatial averages over the REV is omitted.

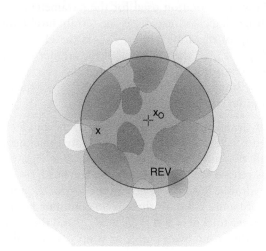

Figure 2.10 Representative elementary volume in a saturated soil (see also colour plate 11)

Source: The authors

Assuming that the grain surface is a material boundary for fluid flow (there is no flow of fluid through the grain surface) and that mass fluxes due to dispersion and diffusion are negligible compared to advective processes, the macroscopic mass balance equation for the fluid, obtained by averaging the microscopic mass conservation equation over the REV, is

$$-\frac{\partial \rho q_i}{\partial x_i} = \frac{\partial \rho n}{\partial t} \qquad (2.3.1)$$

where tensor notation is used for the Cartesian coordinates x_i and the components of any fluid property in their direction (subscript i), ρ is the fluid density, n is the effective porosity of the solid matrix, $q_i = nV_i$ is the specific discharge (often called Darcy velocity), and V_i is the fluid velocity component in the i-th direction averaged over the volume of fluid contained in the REV. For deformable porous media, it is more convenient to express the fluid mass balance using the relative specific discharge (based on the fluid velocity relative to the velocity of the solid matrix, $V_j - V_{sj}$). Assuming that the grains of the solid matrix are microscopically incompressible or, in other words, that the soil deformation is due solely to the rearrangement of grains and the associated change of porosity, we can derive an alternative form of the fluid mass balance equation:

$$-\frac{\partial \rho q_{ri}}{\partial x_i} = n\left(\frac{\partial \rho}{\partial t} + V_{si}\frac{\partial \rho}{\partial x_i}\right) + \frac{\rho}{1-n}\left(\frac{\partial n}{\partial t} + V_{si}\frac{\partial n}{\partial x_i}\right) \qquad (2.3.2)$$

In this equation, the subscript r stands for 'relative' and the subscript s stands for 'solid'. The equation states that the net fluid flow (inflow–outflow) into a control volume, relative to the movement of the skeleton, is stored within the control volume via a change of the fluid density and a change of porosity. A change of porosity is related to macroscopic strain, which, in turn, depends on macroscopic effective stress through the stress–strain relationship. To find the relationship between macroscopic strain and porosity, we first relate the (macroscopic) velocity of the solid phase to the change in porosity via the mass balance equation for the solid phase. Since the grain surface is a material boundary for the solid phase and the solid phase is microscopically incompressible, the net mass flux (inflow−outflow) of the solid phase is stored within a control volume solely as an increase in the volume of the grains. This change in the volume of the grains, in turn, changes the porosity. The corresponding mass balance equation for the solid phase is

$$\frac{\partial V_{si}}{\partial x_i} = \frac{1}{1-n}\left(\frac{\partial n}{\partial t} + V_{si}\frac{\partial n}{\partial x_i}\right) \tag{2.3.3}$$

It provides a relationship between the change of porosity and the movement of the solid matrix, defined by its macroscopic velocity components V_{si}. On the other hand, the divergence of the soil velocity is related to the macroscopic volumetric strain ε of the solid matrix via

$$\frac{\partial V_{si}}{\partial x_i} = \left(\frac{\partial \varepsilon}{\partial t} + V_{si}\frac{\partial \varepsilon}{\partial x_i}\right) \tag{2.3.4}$$

where ε is the first invariant of the strain tensor $\varepsilon = \varepsilon_{ii} = \varepsilon_{11} + \varepsilon_{22} + \varepsilon_{33}$ which in physical terms means the relative growth of the volume with respect to the original volume. Making use of the previous three equations and assuming that deformation of the solid skeleton conforms to

$$\frac{\partial \rho}{\partial t} \gg V_{si}\frac{\partial \rho}{\partial x_i}, \quad \frac{\partial \varepsilon}{\partial t} \gg V_{si}\frac{\partial \varepsilon}{\partial x_i} \tag{2.3.5}$$

the mass balance equation for the fluid simplifies to

$$-\frac{\partial \rho q_{ri}}{\partial x_i} = n\frac{\partial \rho}{\partial t} + \rho\frac{\partial \varepsilon}{\partial t} \tag{2.3.6}$$

It remains to define the constitutive relationships that relate fluid density and macroscopic strain to fluid pressure. In compressible fluids, the change of density is related to the change of pressure as

$$\frac{1}{\rho}\frac{\partial \rho}{\partial p} = \beta, \tag{2.3.7}$$

where β is fluid compressibility. For a fluid saturating a porous material, p is the macroscopic pressure or average pore pressure within the REV. Assuming a macroscopically isotropic and elastic solid skeleton, such that the change in effective stress is due solely to the change in pore pressure, the compressibility of the skeleton can be expressed as

$$\varepsilon = \alpha p,$$
(2.3.8)

where α is the compressibility of the porous skeleton. The previous two relationships are used, along with equations (2.3.1) to (2.3.6), to formulate the final mass balance equation for a compressible fluid in a deformable macroscopically isotropic and elastic porous medium. It takes the form:

$$-\frac{\partial \rho q_{ri}}{\partial x_i} = \rho(n\beta + \alpha)\frac{\partial p}{\partial t}$$
(2.3.9)

To develop a complete mathematical model, this mass balance equation is combined with the macroscopic momentum balance equation. The latter is derived under assumptions, applicable to the majority of practical problems, that the momentum transfer across the solid–fluid interface (i.e. due to drag) is much larger than both the inertial force and the viscous resistance to the flow inside the fluid. In such conditions, the macroscopic momentum equation simplifies to

$$-\left(\frac{\partial p}{\partial x_i} + \rho g\frac{\partial z}{\partial x_i}\right) = n\frac{\mu}{k_{ij}}\left(V_j - V_{sj}\right)$$
(2.3.10)

where z is the vertically upwards coordinate, V_j and V_{sj} are the j-th components of the macroscopic velocity of the fluid and the solid, respectively, μ is fluid viscosity, and k_{ij} is intrinsic permeability. The equation states that the total driving force (= pressure force + gravity) is in balance with the drag (right-hand side (RHS) of the equation). With the assumption of low Reynolds numbers, the drag is parameterized as proportional to the velocity of the fluid relative to the velocity of the solid. From the momentum equation, the relative specific discharge of the fluid q_{rj} is

$$q_{rj} \equiv n(V_j - V_{sj}) = -\frac{k_{ij}}{\mu}\left(\frac{\partial p}{\partial x_i} + \rho g\frac{\partial z}{\partial x_i}\right)$$
(2.3.11)

When this equation is combined with the mass balance equation, it produces the final groundwater flow equation with a single unknown variable: pore pressure p. For the purpose of flow visualization, it is often convenient to use the hydraulic head version of the flow equation. For an incompressible fluid, the potentiometric head (hydraulic head) is defined as

$$H = z + \frac{p}{\rho g}$$
(2.3.12)

and the momentum equation becomes a generalized form of Darcy's law

$$q_{rj} = -K_{ij} \frac{\partial H}{\partial x_i}$$

(2.3.13)

where $K_{ij} = k_{ij} \rho g/\mu$ is the hydraulic conductivity (dimensions LT^{-1}). For a compressible fluid, such that its average density depends solely on pressure, we may use Hubbert's fluid potential

$$H^* = z + \frac{1}{g} \int_{p_0}^{p} \frac{dp}{\rho(p)}$$

(2.3.14)

and obtain an analogous law

$$q_{rj} = -K_{ij} \frac{\partial H^*}{\partial x_i}$$

(2.3.15)

Combining the mass balance and momentum equations yields

$$\frac{\partial}{\partial x_i} \left(\rho K_{ij} \frac{\partial H^*}{\partial x_j} \right) = \rho^2 g(n\beta + \alpha) \frac{\partial H^*}{\partial t}$$

(2.3.16)

Assuming that $|q_{rj} \partial \rho/\partial x_i| \ll n|\partial \rho/\partial t|$ the equation simplifies to

$$\frac{\partial}{\partial x_i} \left(K_{ij} \frac{\partial H^*}{\partial x_j} \right) = S_0 \frac{\partial H^*}{\partial t}$$

(2.3.17)

where $S_0 = \rho g(\alpha + n\beta)$ is usually called the specific storage or specific storativity of a porous medium. It is a function of the compressibilies of both the fluid and the solid matrix, and may be defined as the volume of water added to (or released from) storage per unit volume of porous medium per unit rise (or decline) of the potentiometric head H^*. For convenience, the potentiometric head throughout much of the following text is simply denoted by H.

The basic groundwater flow equation (2.3.17), derived by averaging fundamental equations over the REV, can be used as a basis for three-dimensional mathematical models of groundwater flow. Very often, however, aquifers are geometrically close to two-dimensional surfaces in the sense that their thickness is much smaller than their other spatial dimensions. In such cases, it is often more practical to use two-dimensional models which simulate flow properties averaged over the aquifer thickness and ignore vertical variations. To obtain such a model we need to average the basic flow equation (2.3.17) over the saturated aquifer thickness B, in other words, for a confined aquifer, between the bottom Z_{bot} and the top of the aquifer unit, Z_{top} (Figure 2.9).

Here it is convenient to switch from tensor notation with x_i, V_i, to hydraulic notation, $x = x_1$, $y = x_2$, $z = x_3$, $V_x = V_1$, $V_y = V_2$, $V_z = V_3$, where x, y are the horizontal coordinates and z is the vertical coordinate in the upwards direction. For any general flow property ψ, the depth-averaged value is obtained as

$$\overline{\overline{\psi}}(x, y, t) = \frac{1}{B} \int_{Z_{bot}}^{Z_{top}} \psi(x, y, z, t)\, dz \qquad (2.3.18)$$

Since both Z_{bot} and Z_{top} vary in space and, for unconfined flows, the top of the aquifer (the water table) may vary with time, we have to apply Leibniz rules for the integration of derivatives in finding averages of both temporal and spatial derivatives. For the temporal derivative of a component of a vector the rule is

$$\int_{Z_{bot}}^{Z_{top}} \frac{\partial \psi_i}{\partial t}\, dz = \frac{\partial}{\partial t} \int_{Z_{bot}}^{Z_{top}} \psi_i\, dz - \left(\psi_i \frac{\partial z}{\partial t}\right)\Big|_{Z_{top}} + \left(\psi_i \frac{\partial z}{\partial t}\right)\Big|_{Z_{bot}}, \qquad (2.3.19)$$

i.e.

$$B\overline{\overline{\frac{\partial \psi_i}{\partial t}}} = \frac{\partial B\overline{\overline{\psi_i}}}{\partial t} - \left(\psi_i \frac{\partial z}{\partial t}\right)\Big|_{Z_{top}} + \left(\psi_i \frac{\partial z}{\partial t}\right)\Big|_{Z_{bot}} \qquad (2.3.20)$$

For the spatial derivative of the divergence of a vector, the rule is

$$\int_{Z_{bot}}^{Z_{top}} \frac{\partial \psi_i}{\partial x_i}\, dz = \frac{\partial}{\partial x} \int_{Z_{bot}}^{Z_{top}} \psi_x\, dz + \frac{\partial}{\partial y} \int_{Z_{bot}}^{Z_{top}} \psi_y\, dz - \left(\psi_x \frac{\partial z}{\partial x} + \psi_y \frac{\partial z}{\partial y} - \psi_z\right)\Big|_{Z_{top}}$$
$$+ \left(\psi_x \frac{\partial z}{\partial x} + \psi_y \frac{\partial z}{\partial y} - \psi_z\right)\Big|_{Z_{bot}} \qquad (2.3.21)$$

i.e.

$$B\overline{\overline{\frac{\partial \psi_i}{\partial x_i}}} = \frac{\partial B\overline{\overline{\psi_x}}}{\partial x} + \frac{\partial B\overline{\overline{\psi_y}}}{\partial x} - \left(\psi_x \frac{\partial z}{\partial x} + \psi_y \frac{\partial z}{\partial y} - \psi_z\right)\Big|_{Z_{top}}$$
$$+ \left(\psi_x \frac{\partial z}{\partial x} + \psi_y \frac{\partial z}{\partial y} - \psi_z\right)\Big|_{Z_{bot}} \qquad (2.3.22)$$

or alternatively

$$B\frac{\overline{\overline{\partial \psi_i}}}{\partial x_i} = \frac{\partial B\overline{\overline{\psi}}_x}{\partial x} + \frac{\partial B\overline{\overline{\psi}}_y}{\partial x} + \psi_i|_{Z_{top}}\frac{\partial \left(z - Z_{top}\right)}{\partial x_i} - \psi_i|_{Z_{bot}}\frac{\partial \left(z - Z_{bot}\right)}{\partial x_i} \qquad (2.3.23)$$

By averaging the mass balance equation (2.3.1) over the saturated aquifer thickness and assuming negligible macrodispersion of the total mass $(\overline{\overline{\rho q_i}} \cong \overline{\overline{\rho}}\,\overline{\overline{q}}_i)$ we obtain

$$-\left(\frac{\partial B\overline{\overline{\rho}}\,\overline{\overline{q}}_x}{\partial x} + \frac{\partial B\overline{\overline{\rho}}\,\overline{\overline{q}}_y}{\partial y}\right) = \frac{\partial B\overline{\overline{\rho n}}}{\partial t} - \left(\rho n\frac{\partial z}{\partial t} + \rho q_x\frac{\partial z}{\partial x} + \rho q_y\frac{\partial z}{\partial y} - \rho q_z\right)\bigg|_{Z_{top}}$$

$$+ \left(\rho n\frac{\partial z}{\partial t} + \rho q_x\frac{\partial z}{\partial x} + \rho q_y\frac{\partial z}{\partial y} - \rho q_z\right)\bigg|_{Z_{bot}} \qquad (2.3.24)$$

Alternatively, we can use the equation describing the movement of a surface defined as, for example, $z - Z_{top}(x, y, t) = 0$ at velocity v_i

$$\frac{\partial \left(z - Z_{top}\right)}{\partial t} = -v_i\frac{\partial \left(z - Z_{top}\right)}{\partial x_i} \qquad (2.3.25)$$

to obtain

$$-\left(\frac{\partial B\overline{\overline{\rho}}\,\overline{\overline{q}}_x}{\partial x} + \frac{\partial B\overline{\overline{\rho}}\,\overline{\overline{q}}_y}{\partial y}\right) = \frac{\partial B\overline{\overline{\rho n}}}{\partial t} + \rho n\left(V_i - v_i\right)\big|_{Z_{top}}\frac{\partial \left(z - Z_{top}\right)}{\partial x_i}$$

$$- \rho n\left(V_i - v_i\right)\big|_{Z_{bot}}\frac{\partial \left(z - Z_{bot}\right)}{\partial x_i} \qquad (2.3.26)$$

The last two terms on the right-hand side represent the mass of fluid flowing through the upper and lower boundaries, respectively, per unit time and per unit area of the boundary. In the original three-dimensional mathematical model, these terms were the boundary conditions defined for the upper and lower boundaries of the model. As a result of averaging over the aquifer thickness, in a two-dimensional model they become source terms. In either case, these terms have to be set as known external fluxes or known relationships between external fluxes and hydraulic heads.

The definition of the upper boundary of the aquifer and the associated boundary condition/source term is different for confined and unconfined groundwater flow. In the former case, the upper boundary is the macroscopic boundary of the aquifer, whereas in the latter case it is the free surface of the groundwater. Two cases of particular interest are leaky confined aquifers and phreatic (water table) aquifers.

Leaky confined aquifer

The rate of change of fluid mass within the control volume can be expressed in terms of the potentiometric head by

$$\frac{\partial \overline{\overline{Bn\rho}}}{\partial t} \cong B\frac{\partial \overline{\overline{n\rho}}}{\partial t} \cong \rho BS_0 \frac{\partial H^*}{\partial t},$$

(2.3.27)

where BS_0 is known as aquifer storativity and is denoted by S. Aquifer storativity is dimensionless.

Using the depth-averaged momentum equation and assuming: (1) that the drag exerted by the porous matrix on the fluid is the dominant mechanism of momentum extraction, and (2) that (x, y) are the principal axes of the hydraulic conductivity tensor, we obtain

$$\frac{\partial B\overline{\overline{\rho}}\,\overline{\overline{q}}_x}{\partial x} + \frac{\partial B\overline{\overline{\rho}}\,\overline{\overline{q}}_y}{\partial y} \cong -\frac{\partial}{\partial x}\left(\overline{\overline{\rho}}B\overline{\overline{K}}_x \frac{\partial H^*}{\partial x}\right) - \frac{\partial}{\partial y}\left(\overline{\overline{\rho}}B\overline{\overline{K}}_y \frac{\partial H^*}{\partial y}\right)$$

(2.3.28)

Both the upper and lower boundaries are material surfaces, and are therefore moving at the velocity of the solid, $v_i = u_i$. The boundary conditions in each case are defined in terms of the rates of leakage derived from external sources $\rho\, q_i^{leak}$ ($=$ mass of fluid crossing the boundary per unit time per unit area). For the upper boundary, continuity across the boundary requires that

$$\rho n \left(V_i - u_i\right)\frac{\partial \left(z - Z_{top}\right)}{\partial x_i} = \rho q_i^{leak}\frac{\partial \left(z - Z_{top}\right)}{\partial x_i}$$

(2.3.29)

The condition at the lower boundary is analogous.

In *UGROW*, external sources of leakage are classified as point, linear and areal (distributed). The total leakage rate is g as the sum of the individual leakage rates from each source. The details of how these individual sources are included in the mathematical model are presented at the end of this section.

If the above equations are inserted into the main flow equation (2.3.26), and it is assumed that the variation of density in the x and y directions is negligible, the flow equation for groundwater flow in a leaky confined aquifer becomes

$$\frac{\partial}{\partial x}\left(B\overline{\overline{K}}_x \frac{\partial H^*}{\partial x}\right) + \frac{\partial}{\partial y}\left(B\overline{\overline{K}}_x \frac{\partial H^*}{\partial y}\right) = S\frac{\partial H^*}{\partial t} + q_i^{leak}\Big|_{Z_{top}}\frac{\partial(z - Z_{top})}{\partial x_i}$$
$$- q_i^{leak}\Big|_{Z_{bot}}\frac{\partial(z - Z_{bot})}{\partial x_i}$$

(2.3.30)

As indicated above, the superscript * denoting Hubbert's fluid potential is omitted in the subsequent text.

Unconfined aquifer with a phreatic surface

For an unconfined aquifer, the upper boundary of the control volume is mathematically defined in an identical way to that used for the confined aquifer, that is, by the function $z - Z_{top}(x,y,t) = 0$. However, the boundaries are physically different. In unconfined aquifers, the upper boundary of the control volume is a free surface, not a material surface, and is moving at a different velocity. It is a macroscopic boundary that exists between saturated soil and unsaturated soil with a water content of θ_0. If water entering from external sources is accumulating at the rate N_i, continuity across the free surface requires that

$$\rho n \left(V_i - v_i \right)\big|_{Z_{top}} \frac{\partial \left(z - Z_{top} \right)}{\partial x_i} = \rho \left(N_i - \theta_0 v_i \right) \frac{\partial \left(z - Z_{top} \right)}{\partial x_i} \tag{2.3.31}$$

in other words,

$$\rho n \left(V_i - v_i \right)\big|_{Z_{top}} \frac{\partial \left(z - Z_{top} \right)}{\partial x_i} = \rho N_i \frac{\partial \left(z - Z_{top} \right)}{\partial x_i} + \rho \theta_0 \frac{\partial \left(z - Z_{top} \right)}{\partial t} \tag{2.3.32}$$

Using this condition and the previously defined condition at the bottom boundary we obtain the following mass balance equation

$$-\left(\frac{\partial B\bar{\bar{\rho}}\,\bar{\bar{q}}_x}{\partial x} + \frac{\partial B\bar{\bar{\rho}}\,\bar{\bar{q}}_y}{\partial y} \right) = \frac{\partial B\overline{\rho n}}{\partial t} + \rho N_i \frac{\partial \left(z - Z_{top} \right)}{\partial x_i} + \rho \theta_0 \frac{\partial \left(z - Z_{top} \right)}{\partial t}$$

$$- \rho \, q_i^{leak}\big|_{Z_{bot}} \frac{\partial \left(z - Z_{bot} \right)}{\partial x_i}\, c \tag{2.3.33}$$

where $B = Z_{top} - Z_{bot}$. The free surface level is also the hydraulic head. Thus, $Z_{top} \equiv H$, and $\partial (z - Z_{top})/\partial t = -\partial H/\partial t$. We can evaluate the first term on the RHS (right-hand side) as

$$\frac{\partial B\overline{\rho n}}{\partial t} \cong \frac{\partial}{\partial t} \int_{Z_{bot}}^{H} \rho n \, dz = \int_{Z_{bot}}^{H} \frac{\partial \rho n}{\partial t}\, dz + \rho n\big|_H \frac{\partial H}{\partial t}$$

$$= B\frac{\overline{\partial \rho n}}{\partial t} + \rho n\big|_H \frac{\partial H}{\partial t} \cong \rho n\big|_H \frac{\partial H}{\partial t} \tag{2.3.34}$$

Here, we assumed that $|B\partial \rho n/\partial t| << |\rho n \partial B/\partial t|$. Inserting this relationship into the governing equation it follows that

$$-\left(\frac{\partial (H - Z_{bot})\bar{\bar{\rho}}\,\bar{\bar{q}}_x}{\partial x} + \frac{\partial (H - Z_{bot})\bar{\bar{\rho}}\,\bar{\bar{q}}_y}{\partial y} \right)$$

$$= \rho \left(n - \theta_0 \right)\big|_H \frac{\partial H}{\partial t} + \rho N_i \frac{\partial \left(z - H \right)}{\partial x_i} - \rho q_i^{leak}\big|_{Z_{bot}} \frac{\partial \left(z - Z_{bot} \right)}{\partial x_i} \tag{2.3.35}$$

For water of constant density we obtain

$$
-\left(\frac{\partial \left(H - Z_{bot}\right) \bar{\bar{q}}_x}{\partial x} + \frac{\partial \left(H - Z_{bot}\right) \bar{\bar{q}}_y}{\partial y} \right)
$$

$$
= \left(n - \theta_0\right) \frac{\partial H}{\partial t} + N_i \frac{\partial \left(z - H\right)}{\partial x_i} - q_i^{leak}\Big|_{Z_{bot}} \frac{\partial \left(z - Z_{bot}\right)}{\partial x_i} \tag{2.3.36}
$$

Using equation (2.3.15) to parameterize the depth-averaged fluxes, the equation becomes

$$
\frac{\partial}{\partial x}\left[\left(H - Z_{bot}\right) K_x \frac{\partial H}{\partial x}\right] + \frac{\partial}{\partial y}\left[\left(H - Z_{bot}\right) K_y \frac{\partial H}{\partial y}\right]
$$

$$
= n_{eff} \frac{\partial H}{\partial t} + N_i \frac{\partial \left(z - H\right)}{\partial x_i} - q_i^{leak}\Big|_{Z_{bot}} \frac{\partial \left(z - Z_{bot}\right)}{\partial x_i} \tag{2.3.37}
$$

where $n_{eff} = n - \theta_0$ and is usually called the effective porosity or specific yield. It denotes the volume of water that enters the unsaturated soil above the free surface per unit area, per unit rise of the free surface.

The groundwater flow equations derived for the confined leaky aquifer and the phreatic aquifer include source terms that account for groundwater recharge entering (or leaving) via the top and bottom of the aquifer. They can be further simplified by assuming that all recharge rates have only vertical components, and can be written as

$$
\frac{\partial}{\partial x}\left(T_x \frac{\partial H}{\partial x}\right) + \frac{\partial}{\partial y}\left(T_y \frac{\partial H}{\partial y}\right) + q_z^{bot} - q_z^{top} = S \frac{\partial H}{\partial t} \tag{2.3.38}
$$

In this equation T_x, T_y are aquifer transmissivities in the x and y directions, respectively, and S = aquifer storativity. In turn,

$$
T_x = \begin{cases} K_x \left(Z_{top} - Z_{bot}\right) \\ K_x \left(H - Z_{bot}\right) \end{cases} \quad T_y = \begin{cases} K_y \left(Z_{top} - Z_{bot}\right) & \text{for confined aquifers} \\ K_y \left(H - Z_{bot}\right) & \text{for phreatic aquifers} \end{cases}
$$

$$\tag{2.3.39}$$

$$
S = \begin{cases} S_0 \left(Z_{top} - Z_{bot}\right) & \text{for confined aquifers} \\ n - \theta_0 & \text{for phreatic aquifers} \end{cases} \tag{2.3.40}
$$

The source term q_z^{bot} represents the rate of leakage through the base of the aquifer. The term q_z^{top} represents, for a confined aquifer, the leakage rate through the top aquifer boundary, whereas for a phreatic aquifer it stands for the rate at which water enters storage above the water table. Both q_z^{bot} and q_z^{top} are positive if they are directed upwards, along the z axis. In urban aquifers, a large number of sources, in addition to natural recharge, can contribute to q_z^{top}. Including the contribution from each explicit source is beneficial, because it allows us to specify their recharge rates using the most appropriate models. The following section is devoted to external sources of groundwater recharge.

2.3.3 External sources of recharge

Simulating the relationship between various sources of groundwater recharge and the groundwater flow system is an important function of *UGROW*. For this purpose, the definition of the source terms in the basic flow equation was extended to include a detailed representation of all sources of groundwater recharge entering via the upper boundary.

To perform this analysis, the groundwater flow equation is presented in an integral form, valid for a finite control volume above a plan area $\iint_\Omega dxdy = \iint_\Omega d\Omega$. It takes the form:

$$\iint_\Omega \left[\frac{\partial}{\partial x}\left(T_x \frac{\partial H}{\partial x} \right) + \frac{\partial}{\partial y}\left(T_y \frac{\partial H}{\partial y} \right) + q_z^{bot} - q_z^{top} \right] d\Omega = \iint_\Omega S \frac{\partial H}{\partial t} d\Omega \qquad (2.3.41)$$

As a reminder, the sign convention adopted in *UGROW* for external sources is that recharge from any source is given as positive (+) when it feeds the aquifer, and negative (−) when it drains it.

The sources of recharge through the upper boundary of the aquifer are classified on the basis of their geometry as:

- point sources (e.g. a septic tank or a damaged connection along a water supply main)
- linear sources (e.g. leaking sewers and water supply pipes), and
- areal sources (e.g. release of water from unsaturated soil above shallow unconfined aquifers, and free-draining sanitary landfills).

In general, a single control volume may receive recharge from numerous point, linear and areal sources. To represent this general case, the following data have to be set for each control volume:

- N_p = the number of point sources, Q_{ps} = volumetric recharge rate defined as volume per unit time for each of the sources, (x_s, y_s) = coordinates defining the position of each of the sources $(s = 1,2 \ldots N_p)$
- N_l = the number of linear sources, Q_{ls} = volumetric recharge rate defined as volume per unit length of the source and per unit time for each of the sources, l_s = geometry of the line defining the position of each of the sources $(s = 1,2 \ldots N_l)$
- N_a = the number of areal sources, Q_{as} = volumetric recharge rate defined as volume per unit area of the source per unit time for each of the sources, a_s = geometry of the area defining the position of each of the sources $(s = 1,2 \ldots N_a)$.

Individual sources of all three types are included in the flow equation using Diraq delta functions to mark their position. The total vertical recharge rate through the upper boundary of the aquifer, per unit plan area of the control volume, $dxdy$ is therefore given by:

$$\sum_{s=1}^{N_p} Q_{ps}\delta_{ps} + \sum_{s=1}^{N_l} Q_{ls}\delta_{ls} + \sum_{s=1}^{N_a} Q_{as}\delta_{as} \qquad (2.3.42)$$

The Diraq delta functions δ_{ps}, δ_{ls}, δ_{as} used in this equation are defined as:

$$\iint_{\Omega} Q_{ps}\delta_{ps}d\Omega = Q_{ps}(x_s, y_s) \qquad (2.3.43)$$

$$\iint_{\Omega} Q_{ls}\delta_{ls}d\Omega = \int_{l_s} Q_{ls}dl \qquad (2.3.44)$$

$$\iint_{\Omega} Q_{as}\delta_{as}d\Omega = \iint_{a_s} Q_{as}da \qquad (2.3.45)$$

Summarizing the contributions from all individual sources by using the integral form of the governing flow equation (2.3.41), it is possible to derive a more detailed representation of the recharge rate through the upper groundwater surface:

$$-\iint_{\Omega} q_z^{top} d\Omega = \iint_{\Omega} \left(\sum_{s=1}^{N_p} Q_{ps}\delta_{ps} + \sum_{s=1}^{N_l} Q_{ls}\delta_{ls} + \sum_{s=1}^{N_a} Q_{as}\delta_{as} \right) d\Omega$$

$$= \sum_{s=1}^{N_p} Q_{ps}(x_s, y_s) + \sum_{s=1}^{N_l} \int_{l_s} Q_{ls}dl + \sum_{s=1}^{N_a} \iint_{a_s} Q_{as}da \qquad (2.3.46)$$

The negative sign on the left-hand side of this equation is the result of the adopted sign convention, that is, the flow through the aquifer top boundary in the positive z direction drains the aquifer, and will hence have a negative sign when set as an external source of recharge.

Options for setting recharge rates from recharge sources are various. To obtain a numerical solution of the governing groundwater flow equation, the recharge rates from point, linear and areal external sources are expressed as linear functions of the groundwater hydraulic head H as follows:

- $Q_{ps} = A_{ps}H + B_{ps}$

  where Q_{ps} is the recharge from a point source (dimensions L^3T^{-1}) $\qquad (2.3.47)$

- $Q_{ls} = A_{ls}H + B_{ls}$ $\qquad\qquad\qquad (2.3.48)$

where Q_{ls} is recharge from a linear source along a line l_s contained within a control volume, (dimensions $L^3T^{-1}L^{-1} = L^2T^{-1}$)

- $Q_{as} = A_{as}H + B_{as}$ (2.3.49)

where Q_{as} is the recharge for an areal source over an area a_s contained within the control volume (dimensions $L^3T^{-1}L^{-2} = LT^{-1}$).

The coefficients A and B depend on the assumed physical mechanism of groundwater recharge and may be given as constant values or as values depending on a set of parameters describing a particular source. A constant recharge rate would be defined by setting A to zero.

The governing groundwater flow equation can now be written as

$$\iint_\Omega \left[\frac{\partial}{\partial x}\left(T_x \frac{\partial H}{\partial x}\right) + \frac{\partial}{\partial y}\left(T_y \frac{\partial H}{\partial y}\right)\right] d\Omega + \iint_\Omega q_z^{bot} d\Omega + \sum_{s=1}^{N_p}\left(A_{ps}H + B_{ps}\right)$$
$$+ \sum_{s=1}^{N_l}\int_{l_s}\left(A_{ls}H + B_{ls}\right)dl_s + \sum_{s=1}^{N_a}\iint_{a_s}\left(A_{as}H + B_{as}\right)da_s = \iint_\Omega S\frac{\partial H}{\partial t}\, d\Omega \text{ (2.3.50)}$$

The above relationships between recharge rates and hydraulic head are convenient for the numerical solution, but do not reveal the physical meaning of the parameters A and B. Physical meanings are described below for linear sources of recharge such as leaking pipes because of their special importance for urban groundwater. The reasoning behind point and areal sources of recharge is analogous.

In order to express the coefficients A and B using physically meaningful parameters we have to develop a conceptual model of recharge based on our knowledge of the recharge source. For example, in a hypothetical situation where a detailed field survey produced data on every single crack in a stormwater sewer, it would be possible to simulate flow through each and every crack. In real-world situations, this level of detail is rarely attainable, so the coefficients A and B would normally be obtained by combining field study with model calibration, or simply assumed on the basis of experience. Below, we explain the conceptual models used in *UGROW* to represent urban groundwater recharge from leaking sewers. The models used for leaking water supply mains and the release of water from point and areal sources of recharge are similar.

Figure 2.11 shows typical examples of the relationship between water levels in the aquifer and water levels in a sewer. Also shown are the corresponding equations for the sewer infiltration or exfiltration rate q (inflow/discharge per unit length of the pipe). We first distinguish between infiltration cases (*a*) and (*b*), which occur when the water table is above the sewer, and exfiltration cases (*c*) and (*d*), when the water table is below the sewer.

The infiltration rate depends on the difference between the groundwater level, H, and a representative hydraulic head for the sewer, H_s. For a surcharged sewer (*a*), H_s is the potentiometric level in the sewer. In the free-surface case (*b*), H_s is less clearly defined, since inflow though an individual crack in the pipe depends on the difference between the groundwater level and either the water level in the sewer pipe (for cracks

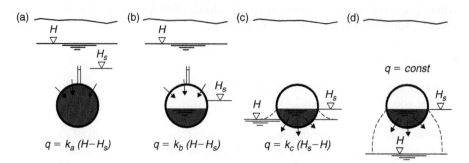

Figure 2.11 Formulas for calculating sewer infiltration rates (a) and (b), and exfiltration rates, (c) and (d), for typical water table conditions

Source: The authors

below the water surface) or the level of the crack (for those above the water level). Since it is not practical to simulate each individual crack, a representative value of H_s is adopted to approximate the bulk influence of all the cracks. It is reasonable to expect H_s to lie between the sewer water level and the crown of the sewer.

For exfiltration, H_s is equal to the hydraulic head in the sewer. For the water table above the pipe invert (Figure 2.11c), the exfiltration rate depends on the difference $H_s - H$, while for deep water tables (Figure 2.11d), it becomes independent of H.

The physical descriptions of these four conceptual models can be translated into corresponding relationships between flow rates and head differences. With infiltration/exfiltration from unpressurized sewers, it is reasonable to expect relatively small flow rates and small corresponding velocities. In such cases, the flow through individual cracks and the surrounding soil can be assumed as laminar so that the resistance law is linear. The corresponding formulas for infiltration/exfiltration rate q (volume per unit time and per unit length of the pipe) are included in Figure 2.11.

To express these formulas in the form $AH + B$ used in the mathematical model, we recall the sign convention: that the flow rate from an external source is positive if it feeds the aquifer and negative if it drains it. With reference to Figure 2.11, the coefficients A and B for the cases (a), (b) and (c) are $A = -k$ and $B = kH_s$ where k is the coefficient evaluated for the particular case (k_a, k_b, k_c for cases (a), (b), (c), respectively). The values of these coefficients depend on the condition of the sewer and the hydraulic conductivity of the surrounding soil. Obviously, the groundwater recharge rate from the external source will be negative in cases (a) and (b) and positive in case (c), which is consistent with the sign convention. In case (d) the coefficients are $A = 0$, $B = q$.

Similar formulas are applicable to water supply pipes. During the normal operation of a water supply system, the pressure head in the pipe is likely to be relatively high. If the water table is below the pipe, the rate of water loss from the pipe is clearly unrelated to the groundwater level. Even for pipes below the water table, the hydraulic head in the pipe is often too high for fluctuations in groundwater level to significantly influence rates of pipe leakage. Thus, during the normal operation of the water supply system, leakage rates are likely to be expressed as values that are independent of the groundwater level (case (d) in Figure 2.11), but still dependent on other parameters such as the state and age of the system. It should be noted that during maintenance and repair, water supply pipes are depressurized, such that those below the water table are likely to receive infiltration from the surrounding aquifer. In such cases, the infiltration rate will clearly depend on the difference between the groundwater level and the head within the pipe (case (*b*) in Figure 2.11). Superimposing groundwater levels on the elevation of the water supply network allows vulnerable parts of the system to be readily detected.

Linear relationships between flow rates and head differences are appropriate for low flow rates. For badly damaged sewers or water supply mains, flow rates may become relatively high and a quadratic resistance law may be more appropriate. In the present version of *UGROW*, a quadratic resistance law for leakage from the external sources is not included as an option.

Analogous conceptual models can be formulated for recharge from point and areal sources. In all cases, they either relate flow rate to the difference between the hydraulic head at the source and the groundwater hydraulic head, or take a known given value for the recharge rate.

2.3.4 Aquifer water balance

The role of the aquifer in the urban water balance is introduced in Section 2.2.2. The basic groundwater flow equation is derived in Section 2.3.2 and conceptual models of the interaction between various external sources and the aquifer are defined in Section 2.3.3. In this section, components of the aquifer water balance are expressed more rigorously.

The integral form of the fundamental flow equation (2.3.50) can be rearranged and transformed, using Gauss's theorem, into

$$-\int_{\Gamma} \left(T_x \frac{\partial H}{\partial x} n_x + T_y \frac{\partial H}{\partial y} n_y \right) d\Gamma + \iint_{\Omega} q_z^{bot} d\Omega + \sum_{s=1}^{N_p} \left(A_{ps} H + B_{ps} \right)$$

$$+ \sum_{s=1}^{N_l} \int_{l_s} \left(A_{ls} H + B_{ls} \right) dl_s + \sum_{s=1}^{N_a} \iint_{a_s} \left(A_{as} H + B_{as} \right) da_s = \iint_{\Omega} S \frac{\partial H}{\partial t} d\Omega \quad (2.3.51)$$

where n_x and n_y are the components of a unit normal vector at the boundary Γ pointing outwards from Ω. The first term on the left-hand side of (2.3.51) represents the net flux (inflow–outflow) through boundary Γ of the area Ω. Other terms on the left-hand side represent various types of external source. In the same order as shown in (2.3.51) these are:

- inflow into the aquifer through the underlying aquitard
- inflow from N_p point sources
- net inflow from N_l linear sources, and
- net inflow from N_a areal sources situated anywhere across the area Ω.

The term on the right-hand side represents the change in storage, that is, the volume of groundwater gained over the area Ω per unit time due to a change in the water table or a change in porosity and water density.

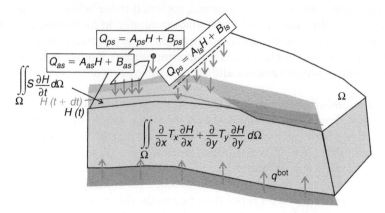

Figure 2.12 Components of the aquifer water balance (see also colour plate 13)

Source: The authors

Individual components of the aquifer water balance are illustrated in Figure 2.12, which shows part of an urban aquifer covering the plan area, Ω, within an enclosed boundary, Γ. The control volume for the water balance extends across the plan area, Ω, and is bounded by the top and bottom of the aquifer. In the unconfined part of the aquifer (left-hand side of the aquifer shown in Figure 2.12) the top of the aquifer is defined by the water table, whereas in the confined part of the aquifer (right-hand side of Figure 2.12) the top of the aquifer is defined by the base of the overlying aquitard. The water balance equation (2.3.51) states that the net flux (inflow–outflow) through the side walls, together with the net contribution from all external sources, is equal to the rate of change of storage within the volume. For an unconfined aquifer, the change of storage per unit time is equal to the volume of water added to the volume between two successive positions of the groundwater table ($H(t)$ and $H(t + dt)$). For the confined aquifer, the change of storage per unit time is equal to the volume of water added to the control volume due to a change in porosity and/or water density.

Equation 2.3.51 presents the aquifer water balance in terms of water volume. A similar balance equation can be generated for chemicals contained in the groundwater, for example, contaminants entering groundwater from an external source. The simplest form of such an equation is obtained if we neglect the hydrodynamic dispersion of the contaminant, and assume that its concentration across the aquifer thickness is constant. For an ideal solute the balance equation is:

$$
-\int_{\Gamma} C_{\Gamma} \left(T_x \frac{\partial H}{\partial x} n_x + T_y \frac{\partial H}{\partial y} n_y \right) d\Gamma + \iint_{\Omega} C^{bot} q_z^{bot} d\Omega
$$

$$
+ \sum_{s=1}^{N_p} C_{ps} \left(A_{ps} H + B_{ps} \right) + \sum_{s=1}^{N_l} \int_{l_s} C_{ls} \left(A_{ls} H + B_{ls} \right) dl_s
$$

$$
+ \sum_{s=1}^{N_a} \iint_{a_s} C_{as} \left(A_{as} H + B_{as} \right) da_s = \iint_{\Omega} CS \frac{\partial H}{\partial t} d\Omega
$$

$$
(2.3.52)
$$

where C is the concentration within the control volume, Ω, and C_Γ, C^{bot}, C_{ps}, C_{ls} and C_{as} are the concentrations of the contaminant in groundwater along the boundary and at various external sources. For positive flux, that is, inflow through the boundary or inflow from an external source, the concentration of the source contaminant has to be known, whereas for a negative flux, the concentration can be assumed to be equal to the concentration in the groundwater at the point where it leaves the aquifer.

2.3.5 Numerical solutions

A two-dimensional mathematical model of flow in an aquifer is derived in Section 2.3.3. The final form of the equation (2.3.50) includes recharge from external sources simulated as linear functions of groundwater hydraulic head. In differential form this equation is:

$$
\begin{aligned}
L(H) &= \frac{\partial}{\partial x}\left(T_x\,\frac{\partial H}{\partial x}\right) + \frac{\partial}{\partial y}\left(T_y\,\frac{\partial H}{\partial y}\right) + \frac{1}{dxdy}\left(A_{bot}H + B_{bot}\right) \\
&\quad + \left[\sum_{s=1}^{N_p}\left(A_{ps}H + B_{ps}\right) + \sum_{s=1}^{N_l}\int_{l_s}\left(A_{ls}H + B_{ls}\right)dl_s + \sum_{s=1}^{N_a}\iint_{a_s}\left(A_{as}H + B_{as}\right)da_s\right] \\
&= S\,\frac{\partial H}{\partial t}
\end{aligned}
\tag{2.3.53}
$$

where $L(H)$ is the function of unknown hydraulic head H (representing the water table in a phreatic aquifer and the potentiometric head in a confined aquifer). Analytical solutions of this equation are available only for special cases involving homogeneous aquifers with a very simple geometry. Unfortunately, the majority of real world engineering problems involve heterogeneous aquifers with an irregular geometry, and numerical methods must be used for solving equation (2.3.53). GROW solves the equation using the finite-element method (FEM). One potential advantage of using FEM as opposed to the finite difference (or finite volume) method (i.e. the method used in MODFLOW) is the ability of the finite-element mesh to represent complex aquifer geometry. The GIS functionality of the UGROW user interface readily handles objects of irregular geometry, so is very well suited for pre-processing and post-processing the finite-element data. Finally, a particular advantage of using FEM is that model development and model calibration are mesh-independent, in other words, the model parameters are assigned to physically meaningful subdomains of the simulation domain rather than to individual model cells.

The modelling procedure starts by defining a conceptual model of the problem. This includes the primary aquifer, the neighbouring hydrogeological units, the simulation domain, the urban water systems, and so on. Once a satisfactory representation of the geometry of the hydrogeological units has been achieved, modelling proceeds by defining the simulation domain (Ω in Figure 2.13), generating of a finite-element mesh covering the domain, defining of all required parameters and running the simulations.

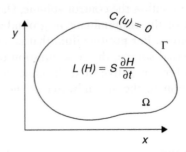

Figure 2.13 Simulation domain, main equation and boundary conditions

Source: The authors

To fully define a problem, the following data are required (a detailed list is presented in Section 2.6):

- Geometry of the aquifer, other urban water system components and the simulation domain.
- Parameters for equation (2.3.53) across the whole simulation domain:
  - aquifer transmissivity T_x, T_y or mean hydraulic conductivity (K_x, K_y), for phreatic aquifers,
  - aquifer storativity S, and
  - parameters describing the recharge rates from each external source: A_{ps} and B_{ps} for point sources; A_{ls} and B_{ls} for linear sources; and A_{as} and B_{as} for areal sources.
- Boundary conditions along the full boundary, Γ, of the flow domain (Figure 2.13). These can be set as the known hydraulic head or recharge rate, or as a linear relationship between the recharge rate and the head. The part of the boundary with known hydraulic head is denoted by Γ_H and the remaining part is denoted by Γ_q. A complete set of boundary conditions can be expressed as

$$C(H) = \begin{cases} H - \bar{H} = 0, & H \in \Gamma_H \\ T_n \dfrac{\partial H}{\partial n} - \bar{q} = T_x \dfrac{\partial H}{\partial x} n_x + T_y \dfrac{\partial H}{\partial y} n_y - \bar{q} = 0, & H \in \Gamma_q \end{cases} \qquad (2.3.54)$$

- Initial conditions, if an unsteady-state simulation is to be run. These comprise values of the hydraulic head at all computational nodes at the beginning of the simulation (at time $t = 0$). Typically, an initial condition is obtained by running a steady-state simulation for conditions relevant for $t = 0$.

Boundary conditions, storativity and the coefficients defining recharge from sources external to the aquifer are generally time-dependent.

Weak formulation of the groundwater flow equation

The finite-element method belongs to the class of weighted residual methods. These search the solutions of integral formulations of the problem, obtained by multiplying both the governing equation (2.3.53) and the boundary conditions (2.3.54) with trial or weighting functions $v, \bar{v}$, and integrating the resulting function over the simulation domain:

$$\iint_{\Omega} vL(H)d\Omega + \int_{\Gamma} \bar{v}C(H)d\Gamma = S\frac{\partial H}{\partial t} \tag{2.3.55}$$

If the solution, H, satisfies (2.3.55) for arbitrarily selected weighting functions it also has to satisfy the original equations (2.3.53) and (2.3.54). The weighting functions are selected in an attempt to minimize the solution error. They appear directly in integrals over Ω so they have to be integrable. The restrictions for the functions representing the unknown solution, H, depend on the highest order of differentiation in $L(H)$. For the problem stated by equation (2.3.53), the solution has to belong to the W_2^2 class, such that the integral of the square of the function itself, and its first and second derivative is limited:

$$\int_{\Omega} \left[H^2 + \left(\frac{\partial H}{\partial x} \right)^2 + \left(\frac{\partial^2 H}{\partial x^2} \right)^2 \right] d\Omega < \infty \tag{2.3.56}$$

The integral formulation is also known as a weak formulation of the problem because the requirements imposed on the choice of functions representing the unknown solution and the weighting functions are weak.

It is reasonable to choose H such that $H - \bar{H} = 0$ along Γ_H, and the weighting functions $\bar{v}$ such that $\bar{v} = 0$ along Γ_H. In this case, the integral or weak formulation of (2.3.53) subject to boundary conditions (2.3.54) can be written as:

$$\iint_{\Omega} v \left[\frac{\partial}{\partial x}\left(T_x \frac{\partial H}{\partial x} \right) + \frac{\partial}{\partial y}\left(T_y \frac{\partial H}{\partial y} \right) + A_{bot}H + B_{bot} \right] d\Omega$$
$$+ \sum_{s=1}^{N_p} v \left(A_{ps}H + B_{ps} \right) + \sum_{s=1}^{N_l} \int_{l_s} v \left(A_{ls}H + B_{ls} \right) dl_s + \sum_{s=1}^{N_a} \iint_{a_s} v \left(A_{as}H + B_{as} \right) da_s$$
$$+ \int_{\Gamma_q} \bar{v} \left[T_x \frac{\partial H}{\partial x}n_x + T_y \frac{\partial H}{\partial y}n_y - \bar{q} \right] d\Gamma = \iint_{\Omega} vS \frac{\partial H}{\partial t} d\Omega \tag{2.3.57}$$

Using Green's formula, the first term in the previous equation becomes

$$\iint_{\Omega} v \frac{\partial}{\partial x}\left(T_x \frac{\partial H}{\partial x} \right) d\Omega = -\iint_{\Omega} \frac{\partial v}{\partial x} T_x \frac{\partial H}{\partial x} d\Omega + \oint_{\Gamma} vT_x \frac{\partial H}{\partial x}n_x d\Gamma \tag{2.3.58}$$

so that the weak formulation of the problem is transformed into

$$
-\iint_{\Omega} \left[\frac{\partial v}{\partial x} T_x \frac{\partial H}{\partial x} + \frac{\partial v}{\partial y} T_y \frac{\partial H}{\partial y} + v A_{bot} H + v B_{bot} \right] d\Omega + \sum_{s=1}^{N_p} v \left(A_{ps} H + B_{ps} \right)
$$

$$
+ \sum_{s=1}^{N_l} v \int_{l_s} \left(A_{ls} H + B_{ls} \right) dl_s + \sum_{s=1}^{N_a} v \iint_{a_s} \left(A_{as} H + B_{as} \right) da_s
$$

$$
+ \oint_{\Gamma} v \left(T_x \frac{\partial H}{\partial x} n_x + T_y \frac{\partial H}{\partial y} n_y \right) d\Gamma + \int_{\Gamma_q} \bar{v} \left(T_x \frac{\partial H}{\partial x} n_x + T_y \frac{\partial H}{\partial y} n_y - \bar{q} \right) d\Gamma
$$

$$
= \iint_{\Omega} v S \frac{\partial H}{\partial t} d\Omega \tag{2.3.59}
$$

The functions v and $\bar{v}$ are chosen arbitrarily, so such functions can be selected such that $\bar{v} = -v$ along Γ_q. The weak formulation becomes

$$
-\iint_{\Omega} \left[\frac{\partial v}{\partial x} T_x \frac{\partial H}{\partial x} + \frac{\partial v}{\partial y} T_y \frac{\partial H}{\partial y} + v A_{bot} H + v B_{bot} \right] d\Omega + \sum_{s=1}^{N_p} v \left(A_{ps} H + B_{ps} \right)
$$

$$
+ \sum_{s=1}^{N_l} v \int_{l_s} \left(A_{ls} H + B_{ls} \right) dl_s + \sum_{s=1}^{N_a} v \iint_{a_s} \left(A_{as} H + B_{as} \right) da_s
$$

$$
+ \oint_{\Gamma - \Gamma_q} v \left(T_x \frac{\partial H}{\partial x} n_x + T_y \frac{\partial H}{\partial y} n_y \right) d\Gamma + \int_{\Gamma_q} v \bar{q} d\Gamma = \iint_{\Omega} v S \frac{\partial H}{\partial t} d\Omega \tag{2.3.60}
$$

We notice that:

- the solution, H, does not appear in the integral along Γ_q, so the flux boundary condition is satisfied automatically, in other words, it is a natural boundary condition,
- if H is selected so that it satisfies the boundary condition in H along Γ_H then v can be zero along that boundary, in which case the integral along $\Gamma_H = \Gamma - \Gamma_q$ vanishes.

The final form of the weak formulation of the groundwater flow equation is thus:

$$
-\iint_{\Omega} \left[\frac{\partial v}{\partial x} T_x \frac{\partial H}{\partial x} + \frac{\partial v}{\partial y} T_y \frac{\partial H}{\partial y} + v A_{bot} H + v B_{bot} \right] d\Omega + \sum_{s=1}^{N_p} v \left(A_{ps} H + B_{ps} \right)
$$

$$
+ \sum_{s=1}^{N_l} v \int_{l_s} \left(A_{ls} H + B_{ls} \right) dl_s + \sum_{s=1}^{N_a} v \iint_{a_s} \left(A_{as} H + B_{as} \right) da_s
$$

$$
+ \int_{\Gamma_q} v \bar{q} d\Gamma = \iint_{\Omega} v S \frac{\partial H}{\partial t} d\Omega \tag{2.3.61}
$$

Finite elements

In the finite-element method (FEM), the flow domain, Ω, is divided into a set of sub-domains called finite elements. Because integrals are additive, each integral over the domain in the weak formulation of a problem can be replaced with sums of the integrals over the subdomains (finite elements). This is achieved by choosing the weighting functions so that these are zero everywhere, except in a finite element. In order to satisfy the integral formulation, the weighting functions must have the following features:

- Their derivatives must be continuous up to the necessary degree. For groundwater flow, where first derivatives are involved in the weak formulation, these are class W_2^1 functions.
- The functions must be continuous at the boundaries between the elements in order to have finite first derivatives. This requirement will be fulfilled if the value of the solution on one side of the element depends only on the values of the solution at the nodes on that side.
- Due to the continuity of the functions and their derivatives, the values of the integrals tend to a constant (are finite) if the area of the FE (the finite element) tends to zero.

The unknown solution, H, is represented as a linear combination of simple basis functions. The simplest interpolation functions are polynomials, so H can be approximated as a sum of polynomials N_i with coefficients a_i. In the Galerkin finite-element method, the weighting functions are selected to be equal to the basis functions:

$$H \approx \sum_1^n a_i N_i, \qquad v_j = N_j, \quad i, j = 1, 2, \dots n \tag{2.3.62}$$

The basis functions (equal weighting functions) are polynomials defined locally, over a single finite element. For this reason it is convenient to use the local coordinates for the basis functions, and define them for a single generic finite element. Figure 2.14 shows a triangular generic finite element with a local coordinate system and the computational nodes at the triangle's edges and mid-sides. The basis functions are therefore the functions of the local coordinates

$$N_i = N_i\left(\xi, \eta\right), \quad i = 1, 2, \cdots n \tag{2.3.63}$$

The number of basis functions n and the order of the polynomial depend on the number of computational nodes. With linear polynomials there are three computational nodes and three polynomials N_i. Each polynomial is non-zero at one computational node and zero at the other two. With quadratic polynomials, the number of nodes and basis functions is $n = 6$. The polynomials are quadratic and non-zero at one computational node and zero at the other five.

Using the rules explained below, the generic finite element, defined in the local coordinate system (ξ, η), is mapped onto each finite element of the simulation domain defined in the global coordinate system (x, y). Thus, the basis functions can also be visualized in the global coordinate system, where each finite element has a set of basis functions which are non-zero at a single node and zero at all other nodes. A single computational node

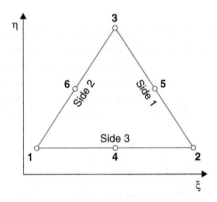

Figure 2.14 A triangular finite element in a local coordinate system

Source: The authors

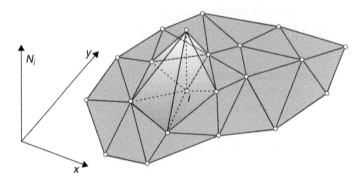

Figure 2.15 Local trial functions in the FEM shown in the global coordinate system over the whole computational domain

Source: The authors

usually belongs to two or more neighbouring elements and hence has an associated basis function in each of them. Figure 2.15 shows an example of linear basis functions associated with a node i in a finite-element mesh made of triangular elements.

Basis functions within a generic finite element are continuous functions of (ξ, η), so the unknown solution, H, is continuous as well. The continuity of the computational domain is achieved by the unique mapping of the sides of the elements from (ξ, η) onto (x, y). The continuity of H from element to element is obtained by the continuity of the domain and the condition that the values of H along a side of an element depend solely on its values at the nodes on that side.

The functions used to map from the local coordinates (ξ, η) onto the global coordinates (x, y) are called shape functions. A special type of finite-element known as isoparametric elements has the same shape function as the trial functions. Any function u inside the element is approximated as

$$u\left(\xi, \eta\right) \approx \sum_{1}^{n} u_i N_i \left(\xi, \eta\right)$$

(2.3.64)

This equation can be used for mapping global coordinates. Global coordinates of a point with local coordinates (ξ, η) are found according to

$$x\left(\xi, \eta\right) \approx \sum_{1}^{n} x_i N_i\left(\xi, \eta\right), \quad y\left(\xi, \eta\right) \approx \sum_{1}^{n} y_i N_i\left(\xi, \eta\right) \tag{2.3.65}$$

where (x_i, y_i) are the coordinates of the computational nodes of the finite element in the global coordinate system. For triangular elements and quadratic basis functions, the number of computational nodes in the element is $n = 6$.

From the relationships $x(\xi, \eta)$, $y(\xi, \eta)$ we can calculate the Jacobian matrix

$$J = \begin{vmatrix} \dfrac{\partial x}{\partial \xi} & \dfrac{\partial y}{\partial \xi} \\ \dfrac{\partial x}{\partial \eta} & \dfrac{\partial y}{\partial \eta} \end{vmatrix} \tag{2.3.66}$$

at any point of a finite element. The matrix can be used to express the partial derivatives with respect to the global coordinates $(\partial/\partial x, \partial/\partial y)$ as a function of the derivatives with respect to the local coordinates $(\partial/\partial \xi, \partial/\partial \eta)$. The relationship is

$$\begin{bmatrix} \dfrac{\partial u}{\partial x} \\ \dfrac{\partial u}{\partial y} \end{bmatrix} = J^{-1} \begin{bmatrix} \dfrac{\partial u}{\partial \xi} \\ \dfrac{\partial u}{\partial \eta} \end{bmatrix} \tag{2.3.67}$$

and can be used to express partial derivatives from the weak formulation (2.3.61), which was derived in global coordinates. The weak formulation contains integrals over finite elements. The integration is carried over elemental areas $d\Omega$ which also have to be mapped from the global coordinate system onto the local coordinate system. The elemental area can be presented as a vector product of the two elemental vectors, $\overrightarrow{d\xi}$ and $\overrightarrow{d\eta}$, which in the global coordinate system have components $\left(\dfrac{\partial x}{\partial \xi} d\xi, \dfrac{\partial y}{\partial \xi} d\xi\right)$ and $\left(\dfrac{\partial x}{\partial \eta} d\eta, \dfrac{\partial y}{\partial \eta} d\eta\right)$, respectively. The elemental area is therefore

$$d\Omega = \left|\overrightarrow{d\xi} \times \overrightarrow{d\eta}\right| = \begin{vmatrix} \vec{i} & \vec{j} & \vec{k} \\ \dfrac{\partial x}{\partial \xi} d\xi & \dfrac{\partial y}{\partial \xi} d\xi & 0 \\ \dfrac{\partial x}{\partial \eta} d\eta & \dfrac{\partial y}{\partial \eta} d\eta & 0 \end{vmatrix} = \left(\dfrac{\partial x}{\partial \xi}\dfrac{\partial y}{\partial \eta} - \dfrac{\partial y}{\partial \xi}\dfrac{\partial x}{\partial \eta}\right) d\xi d\eta = \det J d\xi d\eta \tag{2.3.68}$$

The weak formulation of the problem also contains integrals along the sides of the elements or the lines or areas inside the element. For these integrals, it is sufficient to map a series of individual points (using 2.3.65) along the line (or area) of integration and apply the procedure for numerical integration explained below.

Element matrix

The unknown solution, H, of the weak formulation of the groundwater flow equation *is now approximated* as a linear combination of the basis functions and equal weighting functions. After substituting H in equation (2.3.61) with $\sum_{1}^{n} a_i N_i$ and v with N_j, and changing the order of summing and integration, the following system of equations is obtained:

$$K_{ij} a_i + S_{ij} \frac{\partial a_i}{\partial t} = f_j^q + f_j^B, \quad i, j = 1, 2, \ldots n \tag{2.3.69}$$

where the coefficients K_{ij}, S_{ij}, f_j^q and f_j^B are as follows:

$$K_{ij} = \iint_{FE} \left(\frac{\partial N_i}{\partial x} T_x \frac{\partial N_j}{\partial x} + \frac{\partial N_i}{\partial y} T_y \frac{\partial N_j}{\partial y} \right) d\Omega + \iint_{FE} A_{bot} N_j N_i d\Omega$$

$$- \sum_{s=1}^{N_p} A_{ps} N_j N_i - \sum_{s=1}^{N_l} \int_{l_s} A_{ls} N_j N_i dl_s - \sum_{s=1}^{N_a} \iint_{a_s} A_{as} N_j N_i da_s \tag{2.3.70}$$

$$S_{ij} = \iint_{\Omega} N_j S N_i dx dy \tag{2.3.71}$$

$$f_j^q = \int_{\Gamma_q} N_j \bar{q} d\Gamma_q, \tag{2.3.72}$$

$$f_j^B = B_{bot} N_j + \sum_{s=1}^{N_p} B_{ps} N_j + \sum_{s=1}^{N_l} \int_{l_s} B_{ls} N_j dl_s + \sum_{s=1}^{N_a} \iint_{a_s} B_{as} N_j da_s \tag{2.3.73}$$

In matrix form,

$$[K] = \iint_{FE} [N'][T][N']^T d\Omega + \iint_{FE} A_{bot} [N][N]^T d\Omega$$

$$- \sum_{s=1}^{N_p} A_{ps} [N_{ps}][N_{ps}]^T - \sum_{s=1}^{N_l} \int_{l_s} A_{ls} [N_{ls}][N_{ls}]^T dl_s$$

$$- \sum_{s=1}^{N_a} \iint_{a_s} A_{as} [N_{as}][N_{as}]^T da_s \tag{2.3.74}$$

$$[S] = \iint\limits_{FE} [N] S [N]^T \, dxdy \tag{2.3.75}$$

$$\left[f_i^q\right] = \int\limits_{\Gamma_q} [N] \, \bar{q} d\Gamma_q \tag{2.3.76}$$

$$\left[f_i^B\right] = B_{bot} [N] + \sum_{s=1}^{N_p} B_{ps} \left[N_{ps}\right] + \sum_{s=1}^{N_l} \int\limits_{l_s} B_{ls} \left[N_{ls}\right] dl_s + \sum_{s=1}^{N_a} \iint\limits_{a_s} B_{as} \left[N_{as}\right] da_s \tag{2.3.77}$$

The matrices in the previous four equations are formed as

$$[N] = \begin{bmatrix} N_1 \\ N_2 \\ \dots \\ N_n \end{bmatrix} \quad [N'] = \begin{bmatrix} \dfrac{\partial N_1}{\partial x}, \dfrac{\partial N_2}{\partial x}, \dots, \dfrac{\partial N_n}{\partial x} \\ \dfrac{\partial N_1}{\partial y}, \dfrac{\partial N_2}{\partial y}, \dots, \dfrac{\partial N_n}{\partial y} \end{bmatrix} \quad [T] = \begin{bmatrix} T_x & 0 \\ 0 & T_y \end{bmatrix} \tag{2.3.78}$$

and for external sources, the values of the basis functions are found along their positions

$$\left[N_{ps}\right] = \begin{bmatrix} N_1 \left(\xi_{ps}, \eta_{ps}\right) \\ N_2 \left(\xi_{ps}, \eta_{ps}\right) \\ \dots \\ N_n \left(\xi_{ps}, \eta_{ps}\right) \end{bmatrix} \quad \left[N_{ls}\right] = \begin{bmatrix} N_1 \left(\xi_{ls}, \eta_{ls}\right) \\ N_2 \left(\xi_{ls}, \eta_{ls}\right) \\ \dots \\ N_n \left(\xi_{ls}, \eta_{ls}\right) \end{bmatrix} \quad \left[N_{as}\right] = \begin{bmatrix} N_1 \left(\xi_{as}, \eta_{as}\right) \\ N_2 \left(\xi_{as}, \eta_{as}\right) \\ \dots \\ N_n \left(\xi_{as}, \eta_{as}\right) \end{bmatrix} \tag{2.3.79}$$

The derivatives in $[N']$ are transformed from the global to the local coordinate system to give

$$[N'] = J^{-1} \begin{bmatrix} \dfrac{\partial N_1}{\partial \xi}, \dfrac{\partial N_2}{\partial \xi}, \dots \dfrac{\partial N_n}{\partial \xi} \\ \dfrac{\partial N_1}{\partial \eta}, \dfrac{\partial N_2}{\partial \eta}, \dots \dfrac{\partial N_n}{\partial \eta} \end{bmatrix} \tag{2.3.80}$$

and the elemental area $d\Omega$ is replaced with det $Jd\xi d\eta$ (determinant).

The polynomials N_j are different from zero only in the elements containing the node j, and are equal to zero everywhere else. It is convenient to take the non-zero value of the basis functions N_j as 1, so that the unknown time-dependent coefficients a_i $(H \approx \sum_1^n a_i N_i)$ always take the values of H at the computational nodes of the finite elements. Thus, the solution of the system of equations (2.3.69) will directly produce the time-dependent value of H at each computational node.

Integration over time is performed using an analogous FEM procedure with one-dimensional (over time) linear trial functions. Taking the collocation point to be between two successive time-steps, at $\theta\Delta t$, $(0 \le \theta \le 1)$ the system of equations (2.3.69) can be written as:

$$a_i^{new}\left(\theta K_{ij} + \frac{S_{ij}}{\Delta t}\right) = \overline{f_j^q} + \overline{f_j^B} - a_i^{old}\left[(1-\theta)K_{ij} - \frac{S_{ij}}{\Delta t}\right] \qquad (2.3.81)$$

where a_i^{new} and a_i^{old} are the new and old values of H at the computational node and $\overline{f_j^q}$ and $\overline{f_j^B}$ are the terms on the right-hand side of (2.3.69) evaluated at $\theta\Delta t$.

Solving matrices $K_{ij} S_{ij}$ and vectors $\overline{f_j^q}, \overline{f_j^B}$ in (2.3.70) to (2.3.73) involves integration over areas or along lines. They are usually very difficult or impossible to solve analytically, so numerical integration using Gauss's quadrature formulas is carried out instead. For example, an integral of any function f over a side of the element in the interval $\xi, \in [-1,1]$, $\int_{-1}^{1} f(\xi)d\xi$, (Figure 2.16) is approximated by

$$\int_{-1}^{1} f(\xi)d\xi \cong \sum_{1}^{n} H_i f(\xi_i) \qquad (2.3.82)$$

where ξ_i are Gauss's points and H_i are the corresponding weighting coefficients. An analogous formula is used for integration over the area.

System of equations

The system of linear algebraic equations (2.3.69) is assembled for each finite element. It is represented by a matrix of coefficients and a vector of right-hand-side values. The size of the element matrix is $n \times n$ where n is the number of equations and unknown values, in other words, the number of computational nodes in an element. For a triangular element and quadratic basis functions, $n = 6$.

The matrices of all individual finite elements are assembled into a single global matrix which contains N rows, where N is the total number of computational nodes. The global system is solved using the iterative pre-conditioned bi-conjugate gradient method. The solution contains the values of H at each computational node. The

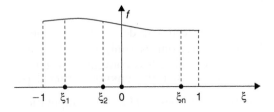

Figure 2.16 Numerical integration along a line

Source: The authors

procedure of assembling the global matrix and solving the global system is repeated for each time-step, until the specified simulation time has been exhausted.

Simulation results can be viewed as 3D pictures of groundwater table/potentiometric surface at various times, as diagrams $H(t)$ at any computational node, or as tables containing the groundwater balance components.

2.4 UNSATURATED SOIL WATER MOVEMENT (*UNSAT*)

2.4.1 Basic equations

The vadose zone includes a solid phase (soil grains, plant roots, man-made material etc.), water and air. The presence of air means that the porous material is unsaturated. In this zone, the fundamental equations governing the movement of water at the microscopic scale (scale of a fluid particle) and their parameters are well known. However, currently available computing resources are insufficient for simulating water movement at this scale in any real-world engineering problem. To overcome this difficulty, the fundamental equations and associated parameters are up-scaled by spatial averaging. The averaging is performed over a representative volume large enough to ensure that the result does not depend on the size of the volume and small enough to exclude the effects of large-scale soil heterogeneities. The volume that satisfies these requirements is the Representative Elementary Volume (REV) (Figure 2.17). The result of averaging is assigned to the centre of the volume denoted by **x** in Figure 2.17. For simplicity, spatially averaged, that is, macroscopic variables are identified only descriptively below, and a special notation for spatial averages over the REV is omitted.

In a system containing more than one phase, separate macroscopic basic equations can be derived for each phase. For soil containing water and air, this means separate equations for the behaviour of water, air and the soil grains. This is simplified by the assumption that the porous matrix is rigid: soil grains do not move, and the air is at atmospheric pressure, in other words, there is no trapped air. Under these assumptions simulating the migration of water alone is sufficient.

Water migration in the unsaturated soil is driven by capillary forces and gravity, such that vertical movement is usually predominant. Neglecting horizontal water

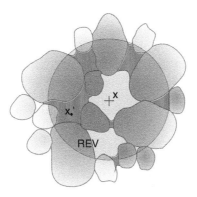

Figure 2.17 Representative elementary volume in unsaturated soil: green = water, grey = air, brown = solids (see also colour plate 12)

Source: The authors

migration further simplifies the modelling process, making the solution one-dimensional. In this case, the continuity equation for the water phase is

$$\frac{\partial \theta}{\partial t} = -\frac{\partial q}{\partial z} \qquad (2.4.1)$$

where z is a vertically *downwards* oriented coordinate, t is time, θ is the water content (volume of water, per unit volume of the control volume Adz) and q is the unit volume flux (volume passing per unit area A per unit time).

The relationship between water flux and the driving force is expressed by a generalized form of Darcy's law:

$$q = -k\left(\frac{\partial h}{\partial z} - 1\right) \qquad (2.4.2)$$

where k is the unsaturated hydraulic conductivity and $h = p/\rho g$ is the pressure head (usually called capillary pressure head), where p is the macroscopic pressure in the unsaturated soil (capillary pressure) and ρ is water density. If we use atmospheric pressure as the datum, such that p is the gauge pressure, capillary pressures in unsaturated soils will typically be negative. The unsaturated hydraulic conductivity k depends on the water content $k = k(\theta)$ because varying water content changes the geometrical configuration of water within the soil which, in effect, changes the soil's pathways for water transmission.

Combining equations (2.4.1) and (2.4.2) gives

$$\frac{\partial \theta}{\partial t} = \frac{\partial}{\partial z}\left[k\left(\theta\right)\left(\frac{\partial h}{\partial z} - 1\right)\right] \qquad (2.4.3)$$

This equation includes two unknowns, θ and h, and a single parameter, k. However, θ and h are related to each other because higher capillary pressure (in absolute terms) is generated when just the small pores become saturated and hence corresponds to low water content, whereas low capillary pressure indicates that the water content is high with the larger pores also filled with water. The shape of the function $h(\theta)$ depends on the history of water migration, that is, it exhibits hysteresis with a notable difference between drainage and wetting. The difference is schematically explained by a so-called ink-bottle effect where the capillary suction at the pore 'necks' hold water in the wet sample during a drainage cycle, while the pore 'bottles' inhibit wetting during a wetting cycle. Thus, for the same suction head, water content is typically higher during drainage than during wetting. In *UNSAT,* hysteresis in $h(\theta)$ is neglected so that the relationship $h(\theta)$ is assumed unique. The relationships $h(\theta)$ and $k(\theta)$ are known as soil characteristics or soil suction curves. Their typical shapes are shown in Figure 2.18.

Since both $h(\theta)$ and $k(\theta)$ are assumed unique we can replace equation (2.4.3) with an equation with either water content θ or capillary pressure head h. Using the pressure head option has the advantage of being able to simulate positive pressures that occur, for example, after ponding of the land surface during a heavy storm when rainfall

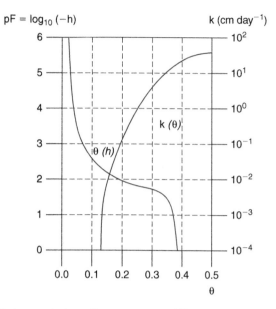

Figure 2.18 Typical soil characteristics: soil moisture curve θ(h) and hydraulic conductivity as a function of moisture content k(θ). Capillary pressure head h is in cm.

Source: The authors

intensity exceeds the infiltration capacity of the soil. Expressing (2.4.3) in terms of h yields the well-known Richards equation:

$$C(h)\frac{\partial h}{\partial t} = \frac{\partial}{\partial z}\left[k(h)\left(\frac{\partial h}{\partial z} - 1\right)\right]$$ (2.4.4)

where C is the soil water capacity and $C(h) = d\theta/dh$.

From the shape of the soil characteristics curves shown in Figure 2.18, it is clear that equation (2.4.4) is non-linear. In *UNSAT*, the soil characteristics are approximated using well-established Van Genuchten relationships (Van Genuchten, 1980).

$$S_e = \left(1 + (\alpha|h|)^n\right)^{-m} \quad h < 0,$$ (2.4.5a)

$$S_e = 1 \quad h > 0,$$ (2.4.5b)

$$\frac{k}{K_s} = S_e^{1/2}\left[1 - (1 - S_e^{1/m})^m\right]^2$$ (2.4.5c)

where α and n are soil parameters and $m = 1 - 1/n$, k is the hydraulic conductivity of the unsaturated soil, K_s is the hydraulic conductivity of the saturated soil and S_e is the relative saturation defined as:

$$S_e = \frac{\theta - \theta_r}{\theta_{max} - \theta_r} \tag{2.4.6}$$

In equation (2.4.6), θ_r is the residual water content and θ_{max} is the maximum water content approximately equal to porosity.

2.4.2 Numerical solution

UNSAT solves the Richards equation for a series of vertical columns of soil extending between the ground surface and the aquifer. For each column, the following has to be defined:

● initial conditions: the capillary pressure head along the whole column at the beginning of the simulation, $h(z,t = 0)$,
● boundary conditions: known capillary pressure head h (or some condition related to h) at the top and the bottom of the column,
● soil parameters: $k(h)$ and $C(h)$,
● computational grid: spatial step(s) Δz and temporal step(s) Δt.

The solution procedure and the data required to solve the Richards equation for a soil column are schematically shown in Figure 2.19. The equation is solved using a Godunov-type finite volume scheme. For this purpose, (2.4.4) is re-written as

$$C(h)\frac{\partial h}{\partial t} = -\frac{\partial q}{\partial z} \tag{2.4.7}$$

The flow domain is split into a series of finite volumes, j, stretching between $z(j - 1/2)$ and $z(j + 1/2)$. Equation (2.4.7) is solved for h within each computational volume with flux evaluated at boundaries between each volume. In any computational time-step, $\Delta t_k = t_{k+1} - t_k$, both the parameter $C(h)$ and the flux between the computational volumes are evaluated at a point in time $k + 1-\varepsilon$, defined by the weighting factor ε. The discretized form of (2.4.7) used in UNSAT is hence:

$$C_j^{k+1-\varepsilon}\frac{h_j^{k+1} - h_j^k}{\Delta t_k} + \frac{2(q_{j+1/2}^{k+1-\varepsilon} - q_{j-1/2}^{k+1-\varepsilon})}{\Delta z_{j-1} + \Delta z_j} = 0 \tag{2.4.8}$$

For $\varepsilon = 1$, the scheme (equation 2.4.8) is explicit; for $\varepsilon = 0$ it is fully implicit. The flux q in equation (2.4.8) is approximated as:

$$q_{j-1/2}^{k+1-\varepsilon} = -k_{j-1/2}^{k+1-\varepsilon}\left(\frac{(1 - \varepsilon) \cdot h_j^{k+1} + \varepsilon \cdot h_j^k - (1 - \varepsilon) \cdot h_{j-1}^{k+1} - \varepsilon \cdot h_{j-1}^k - \Delta z_{j-1}}{\Delta z_{j-1}}\right),$$

$$\tag{2.4.9a}$$

$$q_{j+1/2}^{k+1-\varepsilon} = -k_{j+1/2}^{k+1-\varepsilon} \left(\frac{(1-\varepsilon) \cdot h_{j+1}^{k+1} + \varepsilon \cdot h_{j+1}^{k} - (1-\varepsilon) \cdot h_{j}^{k+1} - \varepsilon \cdot h_{j}^{k} - \Delta z_{j}}{\Delta z_{j}} \right)$$

$$(2.4.9b)$$

Parameters $C(h)$ and $k(h)$ in (2.4.8) and (2.4.9) are evaluated as

$$C_{j}^{k+1-\varepsilon} = C\left(h_{j}^{k+1-\varepsilon}\right) = C\left((1-\varepsilon) \cdot h_{j}^{k+1} + \varepsilon \cdot h_{j}^{k}\right)$$

$$(2.4.10)$$

$$k_{j-1/2}^{k+1-\varepsilon} = k\left(h_{j-1/2}^{k+1-\varepsilon}\right) = k\left((1-\varepsilon) \cdot h_{j-1/2}^{k+1} + \varepsilon \cdot h_{j-1/2}^{k}\right),$$

$$h_{j-1/2} = 0.5\left(h_{j-1} + h_{j}\right)$$

$$(2.4.11)$$

Starting from a known initial condition, the solution involves moving through time using (2.4.8) and (2.4.9) to calculate the capillary pressure head at all computational nodes. If $\varepsilon = 1$, the scheme is explicit such that new values of h are calculated directly.

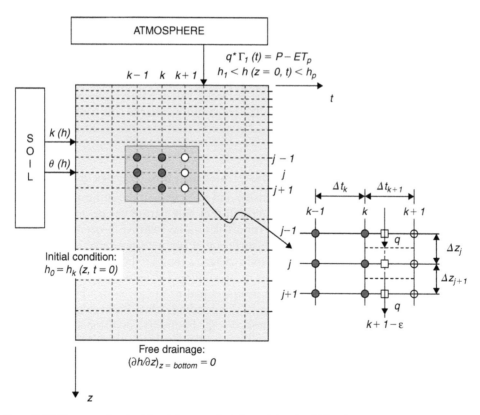

Figure 2.19 Computational grid and data required for solving the Richards equation

Source: The authors

If $\varepsilon \neq 1$, the scheme is implicit and (2.4.8) and (2.4.9) are used to form a system of linear algebraic equations which are solved for h. To complete the solution at each new time-step, the appropriate conditions have to be provided at both domain boundaries, namely at the top and the bottom of the soil zone.

2.4.3 Boundary conditions

The flow domain for water migration in unsaturated soil is one-dimensional, so boundary conditions are required at two points: the land surface and the base of the unsaturated soil zone.

The boundary condition at the soil surface is based on the given precipitation and potential evapotranspiration, which can be either measured and/or evaluated using an empirical formula. Potential evapotranspiration is just a theoretical maximum which may or may not be achieved, depending on the soil and atmospheric conditions. Similarly, depending on the soil characteristics, the precipitation may either infiltrate into the soil at the same rate as the rainfall intensity, or it may pond on the land surface and gradually seep into the soil. Thus the actual evapotranspiration and infiltration from precipitation must be evaluated during the solution, using an iterative procedure. In the first iteration, the boundary condition is specified as known flux (Neuman type boundary condition) equal to:

$$q^*_{\Gamma_1} = P - ET_p \qquad (2.4.12)$$

where Γ_1 is the upper boundary (soil surface), P is precipitation and ET_p is the potential evapotranspiration. The flux $q_{\Gamma 1}$ is a potential value, based solely on meteorological conditions. The actual value depends on conditions in the soil, and its ability to accept or release the calculated amount of water. The solution branches, depending on the sign of flux, as follows.

- Positive $q^*_{\Gamma_1}$. A potential value of flux greater than zero indicates infiltration into the soil. If the precipitation is intensive and greater than the infiltration capacity of the soil, ponding occurs with the result that only part of the precipitation infiltrates, the excess creating surface runoff. Using an iterative procedure, ponding time (i.e. the moment when the soil surface first becomes saturated and a layer of water begins to accumulate on the soil surface) is found to be point at which water saturation at the soil surface reaches its maximum. This maximum corresponds to the value of the capillary pressure head h equal to the given depth of the water layer on the soil surface h_p, known as the ponding value. From this moment on, and as long as the potential flux is positive, the boundary condition at the soil surface is of a Dirichlet type, such that the capillary pressure head h is equal to the given value h_p. This switching of boundary conditions occurs because the water content (and therefore capillary potential) cannot be further increased by infiltration. The value of h_p is usually zero, based on the assumption that runoff starts as soon as the soil is saturated.
- Negative $q^*_{\Gamma_1}$. A potential value of flux less than zero indicates evapotranspiration from the soil. In this case, the actual flux at the soil surface (actual evapotranspiration)

is limited by the ability of the air above the soil surface to maintain the evaporated water in the vapour phase, as well as by the ability of the soil (and vegetation) to transport water from depth to the soil surface. The iterative procedure consists of locating the moment at which the water content at the soil surface reaches equilibrium, corresponding to the humidity and temperature of the atmospheric air. From this moment on, the boundary condition at the soil surface becomes the capillary potential, corresponding to the equilibrium moisture content. In analogy with infiltration, switching of the boundary condition occurs here because, due to the nature of the atmosphere, the moisture content cannot be further reduced by evapotranspiration.

The above procedures can be summarized as follows. The real flux is limited by the infiltration capacity of soil, while the capillary pressure head at the soil surface cannot be larger than the ponding capillary pressure head h_p, nor smaller than the head h_l which corresponds to the equilibrium moisture content:

$$|q^*_{\Gamma 1}| \le |q_{\Gamma 1}| = \left| -k(h)\left(\frac{\partial h}{\partial z} - 1\right) \right|$$

(2.4.13a)

$$h_l \le h_1 \le h_p$$

(2.4.13b)

$$h_l = \frac{RT}{Mg} \ln\left(\frac{RH}{100}\right)$$

(2.4.13c)

In these equations, h_l is the capillary potential at the soil surface (at computational node No.1), R is the universal gas constant [Jmol^{-1} K^{-1}], T is air temperature [K], g is the acceleration due to gravity, M is the molecular weight of water [kg mol^{-1}] and RH is relative humidity [%].

The boundary condition at the base of the soil column depends on the level of the water table. For a deep water table, below the base of the soil zone, this boundary condition is referred to as free drainage, whereby water is allowed to leave the soil column at a rate controlled by its hydraulic conductivity. This boundary condition is a von Neumann type:

$$q_{\Gamma_{Nj}} = -k(h_{Nj})$$

(2.4.14)

where N_j is the number of the computational node at the base of the unsaturated soil. If the depth to the water table, d, is smaller than the thickness of the unsaturated soil column, z_{Nj}, the pressure head at all computational nodes below the water table is known and equals:

$$h_j = z_j - d, \quad \forall j : z_j \ge d$$

(2.4.15)

The lower boundary condition is given at the highest node below the ground-water table. This boundary condition is a Dirichlet type.

2.4.4 Simulation results

Starting from a known initial condition, the solution involves moving through time, calculating all unknown pressure heads, h_j^{k+1}, at each new step, t_{k+1}. At all interior nodes $(1 < j < N_j)$ the discretized form of the basic equation (2.4.8) and (2.4.9) is used to form a set of linear algebraic equations with unknown h_j^{k+1} values. The boundary conditions implemented at the land surface $j = 1$ and the unsaturated soil base $j = N_j$ (or the highest node below the water table) complete the set of algebraic equations. The set of equations has tri-diagonal form and can be solved using any algorithm suitable for such a form. The results of the solution are the values of the unknown capillary pressure head h_j^{k+1} at every nodal point j at the next simulation time t_{k+1}. Once these values are known, fluxes between any two computational nodes can be calculated using the discretized form of the generalized Darcy's law. In the context of *UGROW*, the most important simulation result is the flux at the lower boundary of the unsaturated soil column. Depending on the level of the water table, it is calculated either from the free-drainage boundary condition (2.4.14) or from the gradient of h^{k+1} just above the water table. The flux that leaves the unsaturated soil column through the lower boundary is used in the groundwater simulation as known aquifer recharge.

Moisture content profiles provide a more intuitive means of visualizing simulation results than pressure head profiles. For this purpose, capillary pressure heads are converted to water content using established relationships (equation 2.4.5a). These are presented as the evolution of moisture content profiles with time.

In *UGROW*, the whole area of interest is divided into zones, according to land use and soil characteristics. Simulation of water migration through the soil is performed for each zone. Results of the simulation can be viewed either as an animated soil moisture profile or as a table that lists the water balance terms for each simulation time-step.

2.5 SURFACE RUNOFF (*RUNOFF*)

The surface runoff simulation model, *RUNOFF*, receives input data from the *UNSAT* model. The input data are the quantities of water that do not infiltrate into the soil, but contribute to runoff. *RUNOFF* calculates the directions and travel times for surface runoff across the whole simulation domain. In other words, *RUNOFF* conveys surface runoff to natural urban streams or into the sewage network.

The software system *UGROW* is primarily dedicated to the simulation of urban aquifers and the interaction of groundwater with the urban water infrastructure. With this in mind, a quick and simple simulation model for surface runoff was adopted. The model does not provide details of overland flow, but can predict the volume of runoff water reaching outlet points, thereby facilitating model calibration. This approach is justified by the fact that simulation time-steps for the groundwater model are normally of the order of days, while urban watersheds are rarely large enough to have travel times of the same order.

Steps in the surface runoff simulation are:

- Delineation (dividing the modelling area into sub-areas or sub-catchments, based on the directions of surface runoff from model elements).

- Calculation of slope, length and concentration (travel) time for all model elements between the source of the runoff and the outlet. The outlet is defined as a point within the stream or sewage network where the model will calculate the hydrograph (discharge versus time).
- Summing the hydrographs, appropriately lagged in time, at the outlets.

2.5.1 Delineation

Delineation is the procedure of dividing the catchment area into sub-areas or sub-catchments, so that each sub-catchment is drained to a single channel.

Since the groundwater simulation model is based on triangular finite elements, the delineation algorithm was developed for a TIN (triangular irregular network)-based digital terrain model. The common practice in GIS-based distributed hydrological models is to use GRID-based algorithms to solve such a problem. However, this methodology is not consistent with the overall approach adopted for *UGROW* because it would introduce unnecessary additional complexity to the model.

There are two main versions of the TIN-based delineation algorithm, both of which are implemented in *RUNOFF*: a propagation-based algorithm and an algorithm based on finding the steepest descent pathway in the mesh.

Propagation-based algorithm

The propagation-based algorithm has two versions: segment-based propagation, denoted by D_3, and node-based propagation, D_n. It involves the following procedure:

1. The first step is to determine the elements intersected by channels or pipes from the sewage or stream network. These elements are treated as sink elements in that neighbouring elements must drain into them. The procedure is referred to as 'burning in'.
2. The steps that follow are a slight modification of Prim's algorithm for the shortest spanning tree. This algorithm is well known from graph theory. The modification involves selecting the maximum slope spanning tree instead of the shortest spanning tree, in other words, the element with maximum slope is selected based on all neighbouring elements. The algorithm divides the elements into three sets:
   a) A set of previously assigned elements (initially, this set consists solely of the sink elements obtained by the burn-in procedure of the first step).
   b) a set of adjacent elements that are neighbours of the assigned elements. In the D_3 version of the algorithm, unassigned elements that share a segment with a signed element are candidates for the set of adjacent elements (Figure 2.20), whereas in the D_n version of the algorithm, candidate elements are those that share a node. Another condition that unassigned elements must satisfy to be part of the adjacent set is to have a terrain slope towards the assigned element.
   c) The third set consists of all remaining, that is, neither assigned nor adjacent elements.
3. From the list of adjacent elements, the one with the maximum slope is selected and becomes a member of the set of assigned elements. The set of adjacent elements is then updated and the previous step is repeated.

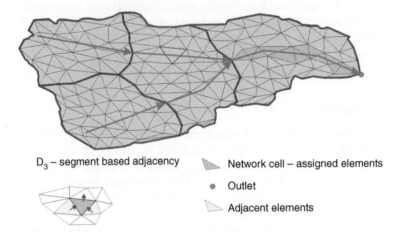

D$_3$ – segment based adjacency   Network cell – assigned elements

Outlet

Adjacent elements

Figure 2.20 TIN-based delineation – D$_3$ propagation algorithm (see also colour plate 14)

Source: The authors

The assignment procedure is repeated until all elements are exhausted, in other words, each element belongs to either set a or set b. In this way the direction of the surface runoff is established for each element, in other words, its neighbouring 'upstream' and 'downstream' elements (regarding the surface runoff) have been identified. The next step is to determine the pathlines for the movement of surface runoff over the elements.

Steepest descent pathway algorithm

The steepest descent version of the TIN-based delineation algorithm uses pathlines passing through the centroid of each element, calculated following the steepest slope (Figure 2.21). For planar triangles, the vertical coordinate z of any point inside the element can be calculated using the following simple plane equation:

$$z = -\frac{A}{C}x - \frac{B}{C}y - \frac{D}{C} \qquad (2.5.1)$$

where A, B, C and D are computed from the coordinates of the three vertices (x_1, y_1, z_1), (x_2, y_2, z_2) and (x_3, y_3, z_3):

$$\begin{aligned}
A &= y_1 \cdot (z_2 - z_3) + y_2 \cdot (z_3 - z_1) + y_3 \cdot (z_1 - z_2) \\
B &= z_1 \cdot (x_2 - x_3) + z_2 \cdot (x_3 - x_1) + z_3 \cdot (x_1 - x_2) \\
C &= x_1 \cdot (y_2 - y_3) + x_2 \cdot (y_3 - y_1) + x_3 \cdot (y_1 - y_2) \\
D &= -A \cdot x_1 - B \cdot y_1 - C \cdot z_1
\end{aligned} \qquad (2.5.2)$$

The direction of the steepest descent is:

$$-\nabla f = \frac{A}{C}\vec{i} + \frac{B}{C}\vec{j} \qquad (2.5.3)$$

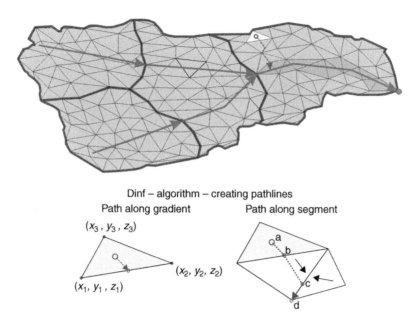

Figure 2.21 TIN-based delineation – D_{inf} algorithm (see also colour plate 15)

Source: The authors

The starting point of a pathline is the centroid of an element (point a in Figure 2.21). The pathline follows the direction of steepest descent until it reaches the boundary of the element (point b). Once the point on the boundary is located, the gradient of the adjacent triangular element is tested to determine if the path continues across that element (points b to c) or along its boundary (as in points c to d).

Computation of the steepest descent pathline continues until the pathline intersects with a drainage channel. From that point the pathline follows the channel through the network until it reaches the outlet.

2.5.2 Time–area diagram and unit hydrograph

The delineation algorithm calculates pathlines from each element (cell) to the outlet. In order to find travel (concentration) times (t_c) from each cell to the outlet, the velocity (V_t) has to be calculated based on the terrain slope (S_t) and the land cover. Following the USDA-SCS procedure, the function $V_t = a S_t^b$ is used, where a and b are coefficients based on the land cover.

The mesh elements can now be classified into time zones (isochrone zones) $j = 1, 2, \ldots$, where each time zone increment is Δt. An element that has a concentration time t_c that satisfies the condition: $(j - 1) \Delta t < t_c < j \Delta t$ belongs to zone j. The time–area diagram is a graph of the cumulative area drained to an outlet within the specified time. It is constructed by summing the incremental areas A_j:

$$A(j\Delta t) = \sum_{k=1}^{j} A_j \tag{2.5.4}$$

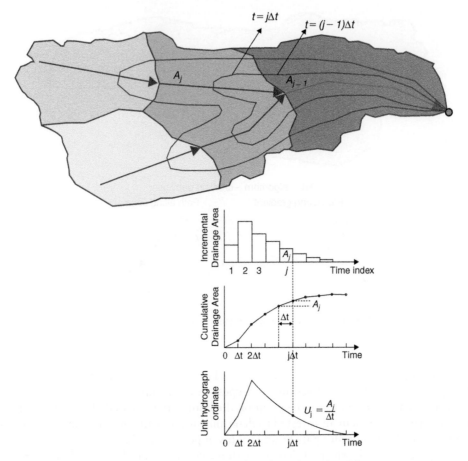

Figure 2.22 Time–area diagram and unit hydrograph (see also colour plate 16)

Source: The authors

From the time–area diagram, the unit hydrograph can be derived (Figure 2.22). The procedure for deriving the unit hydrograph from the time–area diagram is explained by Maidment (1993).

The vertical coordinates of the unit hydrograph are given by:

$$U_j = U(j\Delta t) = \frac{A_j}{\Delta t} \qquad (2.5.5)$$

2.5.3 Direct runoff hydrograph

The portion of rainfall that produces direct runoff (R_{off}) is calculated by the *UNSAT* model, while runoff at the outlet (Q) is calculated from the unit hydrograph.

For example, if the excess rainfall at the first time-step is R_1, then the resulting runoff at the outlet is:

$$Q_1 = R_1 \cdot U_1 = R_1 \cdot \frac{A_1}{\Delta t} \tag{2.5.6}$$

because discharge at the first time-step Δt consists only of runoff from area A_1.

The discharge at the outlet after two time-steps Q_2 is calculated by summing the immediate impact of excess rainfall at time-step 2, R_2 and the delayed effect of R_1 from area A_2.

$$Q_2 = R_1 \cdot U_2 + R_2 \cdot U_1 = R_1 \cdot \frac{A_2}{\Delta t} + R_2 \cdot \frac{A_1}{\Delta t} \tag{2.5.7}$$

Consequently, the discharge at time $t = n\Delta t$ is calculated by summing runoff contributions from each of the available isochrone zones appropriately lagged in time:

$$Q_n = \sum_{j=1}^{n} R_j \cdot U_{n-j+1} = \sum_{j=1}^{n} R_j \cdot \frac{A_{n-j+1}}{\Delta t} \tag{2.5.8}$$

Repeating this procedure over the whole period of simulation produces the hydrograph at the outlet. During model calibration and validation, this hydrograph can be compared with observed data.

2.6 MODEL DATA

UGROW includes a database designed to store data on all features or 'objects' of an urban water system, together with any other data required for model simulation. Section 2.1 (Figure 2.1) covers the database structure, which consists of three major components relating to terrain, geology and water. In this section, we list and describe data contained within each of these components.

Physical systems simulated by *UGROW* consist of a number of 'objects', for example, pipes, aquifer boundaries and areas of land use. Each object has a set of associated data describing its physical properties (such as the length and diameter of a pipe). These are called 'attributes'. A second set of data known as 'properties' defines how each object is graphically presented on the screen (e.g. line thickness and colour). This chapter focuses on the physical system and covers only the attributes of the various objects simulated in *UGROW*.

2.6.1 Terrain

The land surface is represented in *UGROW* as a three-dimensional surface, mathematically described as a digital terrain model (DTM). Input data for a DTM consist of a series of points in three-dimensional space defined by their coordinates (x, y, z) and a series of lines, each of which is defined by a set of points. The points and lines are connected by a surface obtained by spatial interpolation.

To form a DTM, a sufficient number of points and lines needs to be supplied to adequately capture the important features of the land surface. An example of a terrain point is any point on the land surface with known (x, y, z) coordinates, for instance a point read from a contour line on a topographical map. An example of a line is a linear feature that is distinct enough to have to be part of the terrain model. For example, an edge of an excavation could be smoothed by spatial interpolation unless we specify it as a terrain line, which then remains 'fixed' in the DTM. The terrain points are either entered manually by entering their coordinates, or scanned topographical maps of the area are imported and the points along contour lines digitized by clicking on them with the mouse. Terrain points can also be imported from ASCII files containing '*xyz*' extensions. The common '*xyz*' format is simple and contains x, y and z coordinates in each line of the file. Terrain lines are entered by connecting their constituent points either manually or with a mouse. A set of terrain points and lines forms an object called a planar straight line graph (PSLG) or, in simpler terms, a kind of 'cloud' covering the area in question. In *UGROW*, spatial interpolation, that is, forming the DTM, is carried out by triangulation. Triangulation is the process of generating a set of triangles that connect all terrain points, do not overlap, and cover the whole area of interest. The triangulation domain is the region of interest that a user wants to triangulate. It can be convex, in other words, the lines connecting any two points are fully contained within the domain, or not convex when there is a line that is not contained within the domain. Examples of convex and non-convex domains are sketched in Figure 2.23.

In *UGROW* there are two options for triangulation:

- The triangulation domain is fully enclosed within a certain number of lines. In other words, if we connect the boundary lines together, all points and lines are contained within the geometrical shape formed by the boundary lines. In the language of computational geometry, the lines are known as segments and the PSLG is called 'segment-bounded'. The boundary segments enclose the interior of the triangulation domain and clearly separate it from the exterior. In this case, the triangulation domain does not have to be convex.
- The domain of interest is not segment-bounded, so the boundary between its interior and exterior is less clearly defined. In this case, a convex hull is generated over the points and lines of the PSLG. The convex hull is the smallest convex geometrical shape that encloses all points and lines of the PSLG. An example of a convex hull is included in Figure 2.23. Once the convex hull has been generated, the interior of the triangulation domain becomes clearly separated from the exterior.

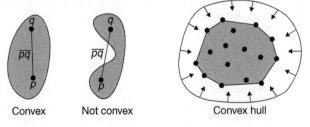

Convex   Not convex   Convex hull

Figure 2.23 Convex triangulation domain and convex hull

Source: The authors

Examples of the two options for triangulation are shown in Figures 2.24 and 2.25. The only difference between the PSLGs shown on the left-hand side of each of these figures is that the first one is segment-bounded, in other words, it has a line connecting boundary points, whereas the second does not include such a line. The outcomes of the triangulation are somewhat different as shown by the right-hand side of each figure. This is because, in the former case, the triangulation domain is not convex, whereas in the latter case, it has to be. Generation of a convex hull in the latter case has changed the shape of the triangulation domain and hence has produced a different set of triangles.

Terrain data are stored in the database in tables as listed in Figure 2.26. The figure also shows the relationships between the tables where the PLSGs are stored. The lines (in fact 'polylines') are stored as sets of segments connecting points. This simplifies the updating of the line geometry and also saves on computer memory because a line does

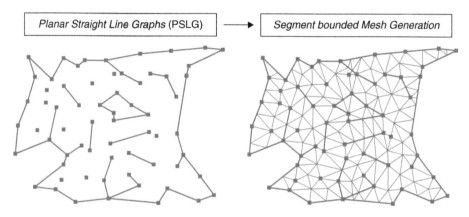

Figure 2.24 **A segment-bounded PSLG**

Source: The authors

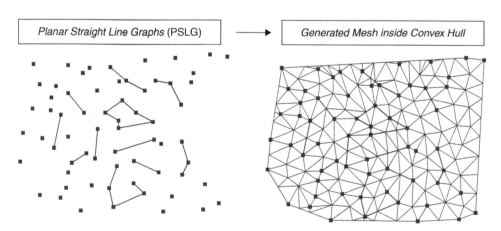

Figure 2.25 **Triangulation inside a convex hull**

Source: The authors

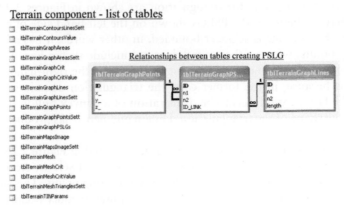

Figure 2.26 Terrain data components

Source: The authors

not have to store its own set of points, and requires only the indices (or addresses) of points from the global set.

Each graph in the database is stored in the same way. The number of such graphs is considerable: the water distribution network, the stream network, the sewerage network, the PSLG defining the groundwater modelling domain, the PSLG containing boreholes, areas defining solid extents and so on.

2.6.2 Geology

The *UGROW* database can store all relevant data on the geological layers present. Mathematically, the layers are represented as three-dimensional objects called solids and have three outer surfaces. The top and bottom surfaces are three-dimensional surfaces spanning the (x, y) plane and are defined as digital models analogous to the digital terrain model. Side surfaces are vertical and formed by vertical lines tracing the shape of the top and bottom surfaces in the (x, y) plane and connecting them.

Data defining the geological layers are set via borehole logs. For each borehole we first define its position by its (x, y, z) coordinates, and then input the borehole log. The log includes data for each geological layer: its name and the elevations $(z,$ coordinates) of its top and bottom. A borehole can be real (and include drilling data) or fictive. The latter type of borehole is normally used to define a geological layer where the geometrical representation of the layer achieved on the basis of real boreholes needs to be improved.

The input of borehole data can be manual; alternatively, digital models of the surfaces between the geological layers can be automatically imported. Once a sufficient number of boreholes is available, layers are formed by creating geology solids bounded by top, bottom and side meshes. The meshes are generated, in turn, by triangulation using the algorithm identical to that used to generate the digital terrain model. There is a special option to ensure that the top surface of a layer never extends above the terrain level, even though this may occur during mesh generation because of imperfect spatial interpolation and the fact that borehole data points are scarce compared to terrain data points. Once all boundary surfaces have been triangulated, the

Geology component - list of tables

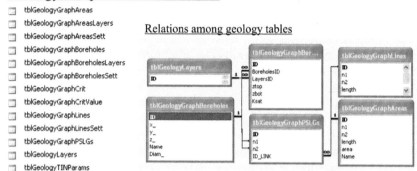

□ tblGeologyGraphAreas
□ tblGeologyGraphAreasLayers
□ tblGeologyGraphAreasSett
□ tblGeologyGraphBoreholes
□ tblGeologyGraphBoreholesLayers
□ tblGeologyGraphBoreholesSett
□ tblGeologyGraphCrit
□ tblGeologyGraphCritValue
□ tblGeologyGraphLines
□ tblGeologyGraphLinesSett
□ tblGeologyGraphPSLGs
□ tblGeologyLayers
□ tblGeologyTINParams

Relations among geology tables

Figure 2.27 **Geology data components**

Source: The authors

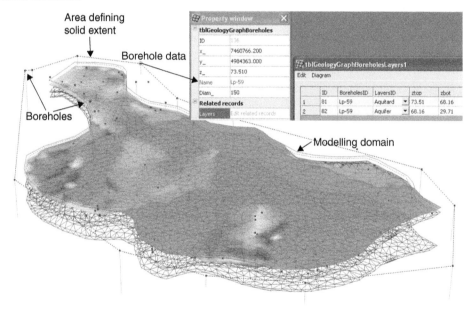

Figure 2.28 **Geological** layers in Pancevacki rit. The details of the case study are presented in
Section 3.2 (see also colour plate 17)

Source: The authors

space between them is subdivided into a series of tetrahedrons through the process of
tessalation. This is a spatial analogue of triangulation and involves generating a series
of tetrahedrons that fill the whole volume of the solid without overlapping.

Figure 2.27 shows the stored geology data components and the relationships
between data tables containing information on boreholes, layers and areas defining
the extent of solids.

Geological layers can be used for general visualization of the geological units and for
defining the aquifer and overlying aquitard for the groundwater simulation model. An
example of a set of geological layers generated using *UGROW* is shown in Figure 2.28.

2.6.3 Water

The third primary component of the database contains data on the various urban water systems, including networks (water supply network, network of sewers and the network of urban streams), the 'soil' vadose zone and the aquifer.

The urban water networks are generally classified into three categories:

- WATNET for water supply pipes
- SEWNET for sewers, and
- STREAMNET for streams.

All three categories contain a set of linear objects (pipes or streams), which together form a network. A pipe always connects two end points defined by their (x, y, z) coordinates. The (x, y) coordinates define the position of the point in the plan view, while z (the pipe level) is the elevation of a selected part of the pipe. This point can be any point in the pipe's cross-section, for example, its centre or base, but it has to be consistent for the whole network. Figures 2.29, 2.30 and 2.31 show single linear objects for each of the networks with a definition of the attributes stored for each object in the database.

WATNET

A water supply pipe (Figure 2.29) has the following attributes:

- name,
- diameter, D,
- length, L,
- reference pressure head, h_0,
- name of the function describing how the pressure head h changes with time,
- reference leakage parameter, k_0,
- name of the function describing how the leakage parameter k changes with time.

The reference pressure head is the average pressure head in the pipe at the point used to define the pipe's level. For example, if the bottom of the pipe is used, as in Figure 2.29, the reference pressure head is evaluated at the bottom of the pipe. During operation of the water supply system, the pressure head is obtained from measured pressures or from the simulation models. During the maintenance or repair of part of a water supply network, the pipe is not pressurized so it becomes similar to a sewer, in other words, the reference pressure head is either zero or very small.

SEWNET

Sewers (Figure 2.30) have the following attributes:

- name,
- diameter, D,
- length, L,

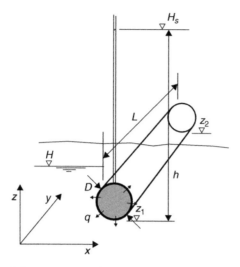

Figure 2.29 Definition sketch for a water supply pipe

Source: The authors

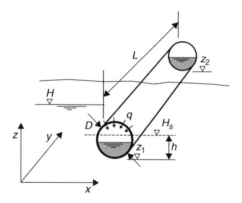

Figure 2.30 Definition sketch for a sewer

Source: The authors

- reference pressure head, h_0,
- name of the function describing how the pressure head h changes with time,
- reference leakage parameter, k_0,
- name of the function describing how the leakage parameter k changes with time.

The reference pressure head is the average pressure head in the pipe at the point used to define the pipe's level. For example, if we use the sewer invert, as in Figure 2.30, the reference pressure head is evaluated at the invert. It may be equal to the water depth in the sewer or it may be a value between the water depth and the full sewer depth

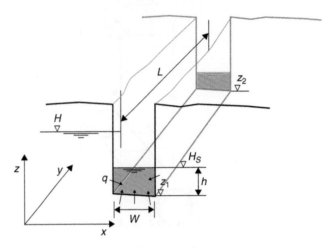

Figure 2.31 Definition sketch for a section of an urban stream

Source: The authors

(as shown in Figure 2.30), which represents the head in the sewer as a consequence of cracks along the whole sewer perimeter.

STREAMNET

A stream (Figure 2.31) has the following attributes:

- name,
- width, W,
- length, L,
- reference pressure head, h_0,
- name of the function describing how the pressure head h changes with time,
- reference leakage parameter, k_0,
- name of the function describing how the leakage parameter k changes with time.

The reference pressure head is the average pressure head in the stream at the point used to define its level. For example, if we use the stream base (as in Figure 2.31), the reference pressure head is evaluated at the base. Typically, the reference pressure head is equal to the depth of water in the stream.

The pressure head and the leakage parameter are used in identical ways for all three types of network. Temporal variations of both the pressure head and the leakage parameter are defined using functions (or 'patterns'), which are a series of non-dimensional values describing the temporal change of any variable. The values of the head or the leakage parameter at any point in time are given by:

$$k(t) = k_0 \times PAT_k(t)$$
$$h(t) = h_0 \times PAT_h(t)$$

(2.6.1)

where $PAT_k(t)$ and $PAT_h(t)$ are the values of the appropriate 'pattern' function at time t. The hydraulic head at a point along a pipe is calculated from the pressure head as

$$H_s(t) = z + h(t) \tag{2.6.2}$$

where z is the pipe level. The hydraulic head, H_s, can be used to calculate the groundwater recharge rate due to leakage from the pipe.

In addition to the attributes assigned to each individual pipe, a common variable is specified that defines the method of calculating the groundwater recharge rate due to pipe leakage. The variable is called 'Leakage Type' and depending on its value, pipe leakage is taken to be either constant or head-dependent. If the 'Leakage Type' variable is set to 2, the recharge is independent of the head and is calculated as

$$Q = k(t) \tag{2.6.3}$$

If the 'Leakage Type' variable is set to 3, the recharge is calculated as

$$Q = \begin{cases} k(t)\left(H_s - H\right) & \text{for } H > H_{min} \\ k(t)\left(H_s - H_{min}\right) & \text{for } H \leq H_{min} \end{cases} \tag{2.6.4}$$

where H_{min} is the lowest level of the water table that still influences the recharge rate. When the water table is lower than H_{min} it is fully decoupled from the source.

Vadose zone

Meteorological data for precipitation, P, and potential evaporation, ET, are required to predict surface runoff and seepage through the unsaturated soil. Temporal variations of P and ET are defined in the same way as for any other time-dependent variable, by a reference value and a 'pattern' function. Thus, reference values P_0 and ET_0 are defined for P and ET, and their values at any moment in time are given by

$$\begin{aligned} P\left(t\right) &= P_0 \times PAT_P\left(t\right) \\ ET\left(t\right) &= ET_0 \times PAT_{ET}\left(t\right) \end{aligned} \tag{2.6.5}$$

Seepage through the soil zone is simulated for parts of the urban area that have a permeable land surface. Often, this is vegetated. In *UGROW* the soil zone is called the 'topsoil'. Areas with identical land use and identical soil characteristics are called 'AreasTopSoil' and are defined in the plan view over the (x, y) plane. The geometrical shape of these areas is constructed from their defined boundary lines. Besides geometrical data, each 'AreasTopSoil' is assigned an appropriate 'topsoil' name. The 'topsoil' name is used for finding the soil parameters required for the calculation of vertical seepage through the unsaturated soil. These parameters include:

- K_z saturated hydraulic conductivity in the vertical direction,
- W_{max} maximum water content, approximately equal to porosity,

- W_r residual water content, and
- α, n-van Genuchten soil parameters which define the soil characteristics.

Groundwater

Data required for the simulations of groundwater flow include fundamental information on the hydrogeological units. Initially, the aquifer and the overlying aquitard must be selected from a list of previously generated geological layers or 'solids'. Subsequent data requirements include:

- The groundwater simulation model domain. The whole domain consists of sub-domains called 'AquiferAreas'. Each of these is created by connecting a set of boundary lines and assigning a name to the material (or rock) it contains. The name of the material can be used as a key for finding the appropriate parameters required. These parameters are:
  - K_x, K_y – hydraulic conductivity in the x and y directions,
  - S_s – specific storativity,
  - S_y – effective porosity (or specific yield) as related to the water table,
  - n – porosity,
  - n_{eff} – effective porosity as related to the pore velocity = the volume of hydraulically active pores/total volume.
- Groundwater simulation model boundaries and boundary conditions. The model boundaries need to be defined in the (x, y) plane as lines created by connecting a series of points. Each boundary is assigned a name, a reference value of the boundary condition ('Head' or 'Flux') and a 'pattern' which describes temporal changes in the boundary condition.
- Finite-element mesh parameters for the aquifer simulation.
- Objects: point and areal sources, wells, etc.

Figure 2.32 shows the relationships between tables in the *GROW* component PSLG that define the modelling domain. The only difference from the *TERRAIN* component PSLG is the nature of the objects derived from the generic point type and line type (e.g. polyline).

2.7 USER INTERFACE

2.7.1 Program overview

3DNet-UGROW (Figure 2.33) is an integrated hydro-informatics tool which contains *TERRAIN*, *GEOLOGY* and *GROW* components, each corresponding to the parts of the physical system simulated by *UGROW*. In a typical application, *3DNet* is used for the step-by-step development of a site-specific model as well as the subsequent running of the model and viewing of the results. The results at any step in the model development and simulation can be viewed as three-dimensional or two-dimensional graphics in the main *3DNet* window. A user communicates with *3DNet* using the 'SceneGraph Window', which lists all objects that can be represented in a *UGROW* application, shows the dialog boxes and provides a toolbar with icons. All these communication methods are described in Section 2.7.2.

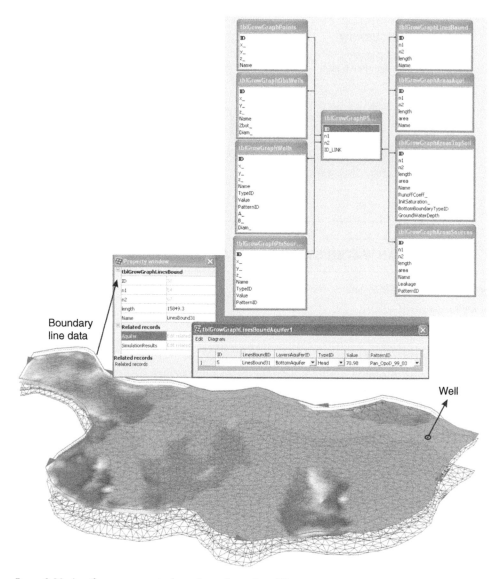

Figure 2.32 Aquifer components (see also colour plate 18)

Source: The authors

All *3DNet* components use computational geometry algorithms for manipulating geometrical data organized into graphical objects, while the *GROW* component contains all the simulation models. All the data are stored in a single external database.

3DNet performs the following tasks:

- connecting to the external database,
- reading 3D graphical objects (GOs) from the database,
- writing GO data to the database,

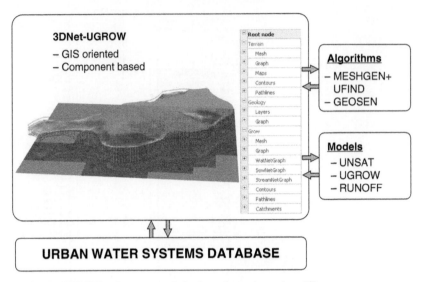

Figure 2.33 3DNet-UGROW and on-screen links (see also colour plate 19)

Source: The authors

- creating 3D and plan views of 'drawing scenes',
- zooming in, zooming out and aligning the viewing centre,
- printing (exporting) drawing scene to files using tiff or dxf graphical formats, and
- cutting the drawing scene with vertical planes.

The *TERRAIN* component is used for:

- inserting and fitting scanned maps,
- inserting (digitizing) elevation points and terrain structural lines,
- triangulating and creating of the Digital Terrain Model (DTM),
- creating contour lines, and
- using predefined or customized colour maps for DTM presentation.

The *GEOLOGY* component is used for:

- inserting real and fictive boreholes,
- defining geological layers by specifying their bottom (base) and top levels in a series of boreholes, as well as their plan area, and
- creating geology layers by forming solids over the study area, between the specified upper and lower boundaries.

The *GROW* component is used for:

- creating the water distribution network (*WATNET*), the urban drainage network (*SEWNET*) and the stream network (*STREAMNET*),

- entering 'topsoil' parameters and other input data for model simulation of unsaturated flow in the vadose zone above the water table,
- running the simulation of unsaturated flow in the vadose zone (this simulation is used for determining aquifer recharge due to precipitation),
- defining hydrogeological units: the aquifer and an overlying aquitard,
- defining the boundaries of the groundwater simulation model domain,
- generating the finite-element mesh for the groundwater simulation,
- determining all sources of groundwater recharge/discharge (e.g. leaking parts of a sewer, recharge from precipitation and leaky water supply pipes), for each finite element, and
- simulating steady and unsteady groundwater flow and displaying results.

A library of tools/algorithms allows for the efficient integration of *UGROW* components. Algorithms currently available include:

- *GEOSGEN* for the generation of geological layers,
- *MESHGEN* for the generation of the finite-element mesh, and
- *UFIND* for assigning sources of recharge to individual elements of the finite-element mesh.

The simulation models include:

- *GROW* for depth-averaged groundwater flow in the main aquifer and vertical flow in overlying and underlying aquitards,
- *UNSAT* for vertical one-dimensional flow in the unsaturated zone above the water table, and
- *RUNOFF* for simulating the surface water balance for the stream and water/sewerage networks.

The basic concepts of GIS (Geographical Information Systems) and object-oriented programming were implemented in the design and development of *UGROW* and its features. These features can be considered from two perspectives:

- The user point of view focuses on the model application and the value of *3DNet* for drawing, inserting, selecting, deleting and updating graphical objects.
- The software design point of view focuses on the internal organization of the data and the functions used for handling them.

In object-oriented programming both data and functions are assigned to objects that belong to abstract classes. The most important abstract class in *3DNet* is called the graphical object (GO). It contains a series of objects.

From the design point of view, there are simple and composite graphical objects. The simple graphical objects are points, polylines and areas (closed polylines). A composite graphical object consists of a series of simple objects and their relationships. A composite GO is implemented as a graph. The simplest graph structure is the Planar Straight Line Graph (PSLG) which is simply a set of points and lines. Special types of

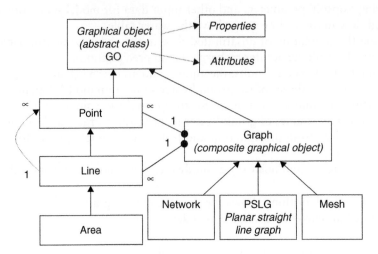

graphs are created for the representation of urban water systems, for example, the urban water network (non-oriented graphs), the stream and sewage networks (oriented graphs), the finite-element mesh and so on.

At the most generic level, all simple or composite graphical objects are derived from the same abstract class: the graphical object (GO). Each object has its own *properties* which define its appearance on the screen (colour, size, etc.) and *attributes* containing the data related to that GO. A sketch showing these basic relationships is provided in Figure 2.34. All urban water system elements are presented as graphical objects. For example, a pipe of a water supply network is a line which has the attributes of length, diameter, age and so on, and the properties of colour, line width, line type and so on.

In accordance with the adopted object-oriented approach, *UGROW* is designed as a 3D graphical engine that has basic drawing and GIS functionalities. The object structure created in this way is known as a Directed Acyclic Graph (DAG) or SceneGraph. Figure 2.34 shows a DAG developed for *UGROW*.

At the top of the *DAG* tree structure, there is a root (Parent) node which controls the limits of the 3D drawing scene. Beneath the root node are the main components of *UGROW – TERRAIN, GEOLOGY* and *GROW –* and these, in turn, contain other composite objects such as tabs.

2.7.2 General 3DNet functions

Work with *UGROW* starts with the creation of a database for the project. Using the File menu commands, a user can create a new database (File ▶ New command) or open an existing one (File ▶ Open command). If a new database is chosen, the program opens a template database which should be saved under a new, unique name using the File ▶ Save As command.

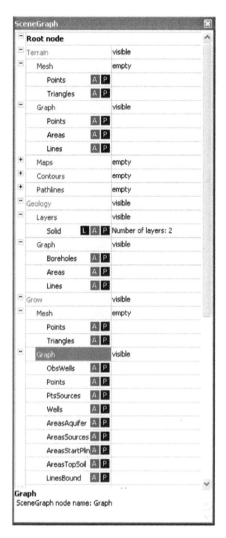

Figure 2.35 UGROW **SceneGraph** *window*

Source: The authors

SceneGraph window

Once the database is loaded, the screen looks similar to the window shown in Figure 2.35. On the left-hand side of the application window is the SceneGraph window, showing the overall structure of the *UGROW* drawing scene.

You can close the SceneGraph window by clicking the X button in its top right corner. It can be redisplayed by clicking its button on the toolbar (it is the only depressed toolbar button in Figure 2.35). Toggling the SceneGraph tool button successively displays and closes the SceneGraph window.

The top of the SceneGraph structure contains the Root node, which controls the contents and the appearance of the scene. The three branches leaving the Root node

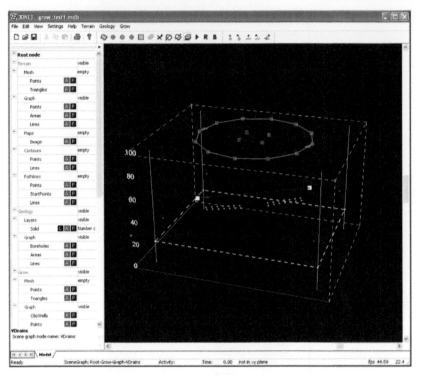

Figure 2.36 Starting layout of *UGROW* interface with the **SceneGraph** window (see also colour plate 20)

Source: The authors

are the three key *UGROW* components: **Terrain**, **Geology** and **Grow**. These components contain other relevant objects such as sub-branches, which are further subdivided into another level of simple objects.

When a node is selected, the corresponding object type becomes active and its name appears at the bottom of the SceneGraph (Figure 2.36). At the same time, the status bar displays the complete 'path' of the node starting from the root, so that the user always knows which node is active if SceneGraph is closed.

Clicking the letter **P** on any node on the lowest level of the SceneGraph structure opens an appropriate property dialog box for the selected node, that is, for the selected type of object. The user can then edit the properties of the objects of the selected type. Choosing the letter **A** shows default attributes associated with the selected object type. The use of dialog boxes is explained below.

Dialog boxes

Selecting menu commands and object types from the SceneGraph window invokes dialog boxes (Figure 2.38), which are used for entering and editing the values of various object properties and attributes. A dialog box provides a record from the database table, with field names in the left-hand column and field values in the right-hand column.

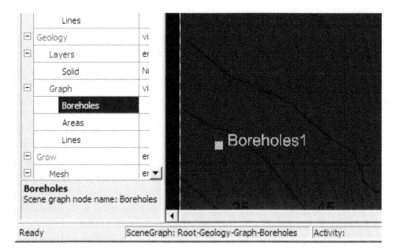

Figure 2.37 The name of a selected node (object type) appears both at the bottom of the
SceneGraph and on the status bar

Source: The authors

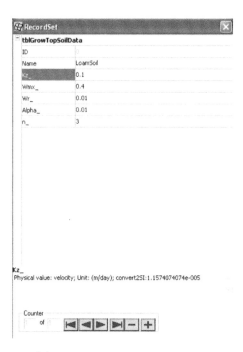

Figure 2.38 Common property dialog

Source: The authors

The user can perform the following operations in a dialog box:

- Change text and numerical values by clicking on a cell containing the value to be changed in order to enter editing mode. Editing mode is indicated by the depressed cell and the blinking cursor.
- Exit editing mode by clicking outside the cell.
- Close the property dialog by clicking the X button in its top right corner.

If a field value is associated with some physical value, information regarding the physical unit is shown at the bottom of the dialog box. Units used by *UGROW* are SI units, so values in the database are always stored in SI. Just before a value appears on the screen it is converted into the appropriate, user-defined unit. Physical units can be changed by selecting the field name and clicking the right mouse button to open the field definition dialog box. In the field definition dialog box, it is possible to change the physical unit and the numerical format of the selected field.

Some of the dialog boxes have a record counter with navigation buttons at the bottom of the window (such as that shown in Figure 2.39). These are related to object types requiring more than one record (e.g. there is usually more than one geological layer, and information for each layer is contained on one record). The user can:

- Add a new record, by clicking the + button.
- Delete a record, by clicking the – button.
- Navigate through the records by using the arrow buttons. The function of these buttons is, from left to right: first, previous, next and last record.

Figure 2.39 Setting the physical unit and precision for a selected field using the field definition dialog

Source: The authors

Handling graphical objects

When an object type is selected in SceneGraph, the user can add or edit particular objects of that type using the drawing tools located on the right-hand side of the tool-bar (Figure 2.40). From left to right, these tools are: Add Points, Edit Points, Add Polylines, Edit Polylines and Drawing Criteria.

Adding points

New points are created as follows:

1. Click on the Add Points tool button 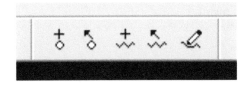 to enable point objects that are active in the SceneGraph to be added. The property window appears (the dialog box to the left in Figure 2.41). This window is the same for all point-like objects (elevation points, boreholes, wells, drainage network nodes, etc.). Note that the 'Add Points' activity is displayed on the status bar. The added point objects first acquire default attributes which can be changed by pressing the letter **A** in the SceneGraph window.

Figure 2.40 **Drawing tools**

Source: The authors

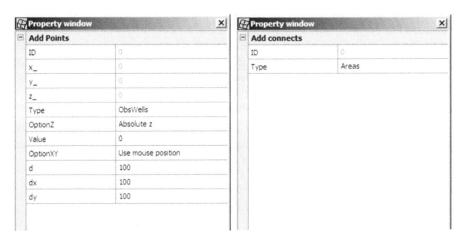

Figure 2.41 **Dialog boxes for adding points, (left-hand window), and polylines (lines and areas), (right-hand window)**

Source: The authors

2. Choose the method for entering X and Y coordinates in the **OptionXY** field of the property window. The following methods are available:
    - Use mouse position: X and Y coordinates are determined from the position of the mouse pointer at the moment the point is clicked.
    - Use exact distance d: the distance between the two points is set by entering the distance in the d field.
    - Use exact relative distance dx & dy: increments in the X and Y directions are set in the dx and dy fields.

    If it is necessary to enter exact X and Y coordinates, the mouse position option can still be used to approximately locate the point, and the coordinates can be revised later in the attribute window, as explained below.
3. Choose the type of information contained in the third coordinate Z in the **OptionZ** field of the property window. The following options are available:
    - absolute z
    - relative z
    - slope %.
4. Enter the value for the Z coordinate in the **Value** field.
5. Add all points with the specified Z value by clicking the mouse.

Adding points is completed by right-clicking the mouse and choosing the **End activity** option from the pop-up menu.

Adding lines and areas

Lines and areas essentially belong to the same type of objects known as *polylines*, except that areas are closed polylines. Polylines consist of points and the links that connect them. Links can be oriented, in other words, they can have direction. To create a polyline, the points must first be entered using the procedure described above. Subsequently, the procedure is to:

1. Click on the **Add Polylines** tool button ⌁ (Figure 2.40). Line objects that are active in the SceneGraph are added during this action. The property window appears (the dialog box displayed to the right in Figure 2.41) and the 'Add Polylines' activity is displayed on the status bar. The added polyline objects acquire default attributes which can be changed by clicking on the letter **A** in the SceneGraph window.
2. Click on the points to be connected (linked) to form the polyline.

Creating a polyline is terminated by pressing the **INS** key. Adding polylines is completed by right-clicking and choosing **End activity** from the pop-up menu.

Editing points

Point-like objects are edited by:

1. Entering edit mode by pressing the **Edit Points** tool button ⬹ (Figure 2.40). The Edit Point Objects dialog is displayed and the 'Edit Points' activity is shown on the

status bar. Only point objects that are active in SceneGraph can be edited during this action.

2. Selecting point(s) by means of one of the following:
   • Click a point on the screen.
   • Drag the mouse around one or more points on the screen.
   • Choose the Attributes in grid option from the pop-up menu to open the grid dialog (Figure 2.42), and click on the appropriate grid line. The Add selection command can be invoked by right-clicking the record counter. After closing the grid dialog, the selected point object will be at the centre of the screen.
   • Note that the easiest way to find a specific object in the grid is to sort according to a selected Attribute value (field value). This is achieved by selecting the field from the header row and right-clicking to sort the table in ascending or descending order. The number of selected objects appears on the status bar of the main window.

3. Starting the desired action. Selected point(s) can be:
   • Deleted by pressing the DEL key.
   • Deselected by pressing the ESC key.

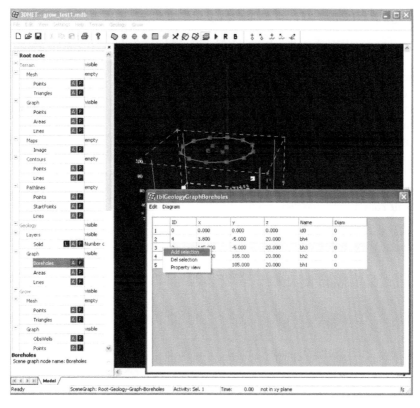

Figure 2.42 **Selecting objects from the grid attribute dialog box (see also colour plate 21)**

Source: The authors

- Modified by right-clicking anywhere on the screen to reveal the pop-up menu which provides options to view or change object Properties and Attributes. Properties define the appearance of the object on the screen and Attributes are data associated with the selected object (in this case, X, Y, Z coordinates). It is also possible to choose the Attributes in grid option to display a spreadsheet table with X, Y, Z coordinates for all the points. In this table, rows with the selected points are coloured yellow.

Edit mode is exited by right-clicking and choosing End activity from the pop-up menu.

Editing polylines

Editing polylines involves essentially the same procedure as that for editing points. The edit mode for polylines is opened by clicking on the Edit Polylines tool button (Figure 2.40).

File menu commands

The following standard 'Windows'-style commands are available in the 'File' menu:

- New: opens a template database which should be saved under a new, unique name using the Save As command.
- Open: opens the existing database.
- Save, Save As: saves changes to the database.
- Print, Print Preview, Print Setup: set printing options and print the drawing scene.
- Write to tiff, Write to dxf: export the drawing scene to TIFF and DXF formats. Files exported in the DXF format are organized in layers that have the same names as the nodes in SceneGraph. Only visible objects are exported.

View menu commands and View tools

The appearance of the drawing scene is controlled by a series of View menu commands or matching View tools:

- zooming options: Zoom All,   Zoom In,   Zoom Out   and Zoom Centre.

- 2D view   and 3D view

- vertical Sections (left and right cutting planes)

- toggle SceneGraph window on and off
- R regenerating Graphical objects
- B updating Graphical objects bounds.

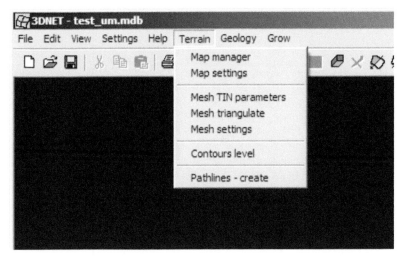

Figure 2.43 **Terrain menu**

Source: The authors

2.7.3 *TERRAIN* component

The *TERRAIN* component controls all tasks related to the preparation of topographic data. The user can:

- insert and fit scanned maps,
- digitize elevation points and structural lines from maps, or import them from another source,
- create the DTM (Digital Terrain Model),
- create contour lines and pathlines,
- view the DTM using predefined or custom colour coding.

These tasks are completed by using commands from the Terrain menu (Figure 2.43).

Inserting scanned maps

A scanned map is inserted using the following procedure:

1. Choose Terrain ▶ Map manager to open the Map manager dialog (Figure 2.44).
2. Click the + button at the bottom of the dialog box to add a new record.
3. Click the Path property field to specify an image file.
4. Define the size and position of the image by entering the lower left (xBL, yBL) and upper right (xTR, yTR) coordinates in the appropriate fields.
5. Check the visible box to show the image on the screen.

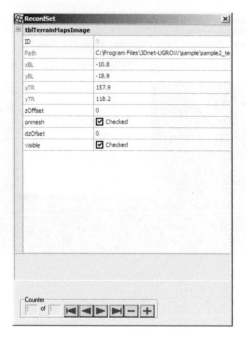

Figure 2.44 Terrain ▶ Map manager dialog

Source: The authors

Digitizing elevation points and structural lines from maps

Elevation points can be digitized from a scanned map by means of the following:

1. In the SceneGraph window select the Terrain-Graph-Points node.
2. Follow the procedure for adding points: press the **Add Points** button on the tool-bar, enter the elevation in the **Value** field of the dialog box, and start adding points with specified elevations by clicking on them (usually by following the corresponding terrain contour).
3. Finish by selecting **End activity** from the right-click pop-up menu.
4. Elevation points can also be added directly in the database table **tblTerrainGraphPoints** or from a simple *.xyz* ASCII file.
5. To add structural lines (lines that will force the digital terrain model generator to Create sides of triangular elements along them) the user should:

1. Select the Terrain-Graph-Lines node in the SceneGraph window.
2. Create the lines from existing points following the procedure described in the section 'Adding lines and areas'.

Creating terrain mesh (Digital Terrain Model)

The Digital Terrain Model, or DTM, is created as a Triangulated Irregular Network, or TIN, which is a mesh consisting of triangles. To create a mesh from elevation points

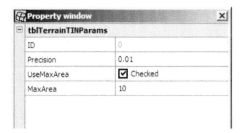

Figure 2.45 Parameters for triangulation of terrain data

Source: The authors

and structural lines already added to the terrain component:

1. Choose the Terrain ▶ Mesh TIN parameters command to open the dialog box.
2. Enter the smallest distance between two points in the **Precision** field (an error will occur if there are two points with the same x, y coordinates).
3. Enter the maximum area (in square metres) of triangles in the mesh in the **MaxArea** field.
4. Start the Terrain ▶ Mesh triangulate command to create the DTM.

To display the TIN mesh, its visibility must be turned on by choosing the **Terrain-Mesh-Triangles** node in the SceneGraph window to display the property window (Figure 2.45), and checking the **Visible** check box. An example of a DTM is presented in Figure 2.46. This DTM is displayed using the properties for mesh triangles shown in Figure 2.47.

2.7.4 *GEOLOGY* component

The purpose of this component is to define the types of geological layer and the layer geometry. Within this component, the user can:

● define the geological layers,
● define both actual and fictive boreholes,
● create geological solids.

The above tasks are performed using the commands listed in the Geology menu (Figure 2.48).

Defining geology layers

At the beginning, you create geological layers by simply assigning them a name and a set of properties. Their position in plan and their elevations are defined later by defining

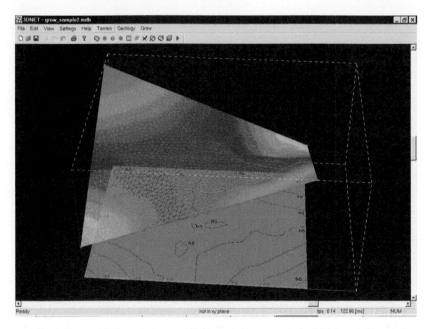

Figure 2.46 An example of DTM created with Terrain ▶ Mesh triangulate command (see also colour plate 22)

Source: The authors

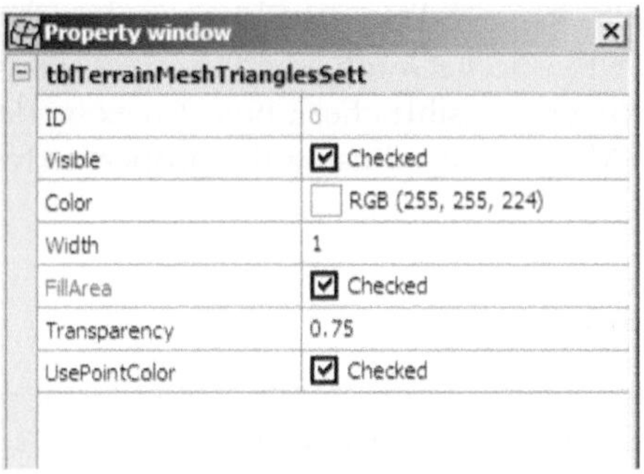

Figure 2.47 Display settings for terrain mesh triangles

Source: The authors

boreholes and assigning layers based on these boreholes. To create geological layers the user should:

1. Run Geology ▶ Layer manager to open the Layer manager dialog box (Figure 2.49).
2. Use the navigation buttons at the bottom of the dialog box to add new layers, delete existing layers, or to select the next or previous layer.
3. Use the options in the dialog box to define the way the layers appear on the screen (colour, line width, visibility etc.) and to give names to the layers.

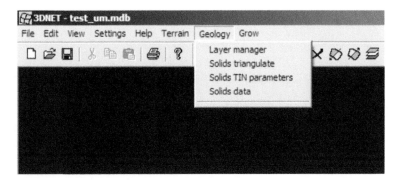

Figure 2.48 **Geology menu commands**

Source: The authors

Defining boreholes

The boreholes are added using the procedure for adding point objects. The user should select the Geology-Graph-Boreholes node in the SceneGraph window and follow the procedure for adding points explained in Section 2.7.2. This procedure positions the boreholes within the coordinate system.

The second step is to define the layers within each borehole. It is necessary to specify the top and bottom elevations for each layer present in a borehole using one of two ways:

- via the Edit Points tool and Attributes dialog
- via the Geology ▶ Layer manager command.

To assign layers to boreholes via the Edit Points tool and Attributes dialog box, the user should:

1. Select the Geology-Graph-Boreholes node selected and press the Edit Points tool button.
2. Select a single borehole on the screen. Right-click and choose Attributes from the pop-up menu. A property window is displayed where the name and the diameter of the borehole can be specified.
3. Click the + sign on the left of Related records and choose the Layers field. A dialog box with a grid is shown.
4. Right-click on the leftmost grid cell and choose Add Record to add a layer to the borehole (Figure 2.50).
5. Select the layer type in the LayersID field and enter the top and bottom elevations of the layer in the ztop and zbot fields.
6. Repeat for each borehole layer.
7. Repeat the procedure for the other boreholes.

Figure 2.49 Geology ▶ Layer manager dialog

Source: The authors

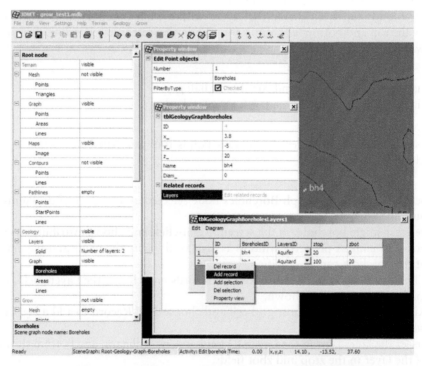

Figure 2.50 Assigning layers to boreholes via the Edit Point tool and Attributes dialog box (see also colour plate 23)

Source: The authors

Layers are assigned to boreholes via the Geology ▶ Layer manager command by performing the following steps:

1. Open the Layer Manager (Geology ▶ Layer manager).
2. Use the navigation buttons at the bottom of the dialog box to select one of the geology layers.
3. Click the + sign on the left of Related records and choose the Boreholes field. A dialog box with a grid is shown.
4. Right-click on the leftmost grid cell and choose **Add record** to add a borehole containing the selected layer (Figure 2.51).
5. Select a borehole from the drop-down list in the **BoreholesID** field and enter the top and bottom elevations of the selected layer in the **ztop** and **zbot** fields.
6. Repeat the above step for each borehole.
7. Repeat the procedure for the other layers.

Creating geological solids

Data on geology layers in the borehole records are used to create solids (bodies) for selected layers.

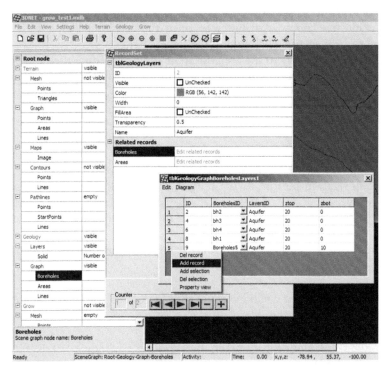

Figure 2.51 Assigning layers to boreholes via the Geology ▶ Layer manager command (see also colour plate 24)

Source: The authors

The first step in this process is to define the plan area, that is, the horizontal extent of the geological solid across the (x, y) plane. For this purpose the user should:

1. Select the Geology-Graph-Areas node from the SceneGraph window. In the property window the Visible box should be checked.
2. Press the Add Polylines tool button. In the Type field of the property window select Areas.
3. Successively select point objects (boreholes) on the screen and finish creating an area by pressing the INS key. To exit adding lines, right-click and choose End activity from the pop-up menu.

In the next step, layers should be assigned to the area created. This procedure is the same as assigning layers to boreholes. Two alternative ways are:

*   via the Edit Lines tool and Attributes dialog box, or
*   via the Geology ▶ Layer manager command.

Layers are assigned to geology areas via the Edit Lines tool and Attributes dialog box as follows:

1. With the Geology-Graph-Areas node selected, press the **Edit Lines** tool button and select an area on the screen.
2. Right-click and choose Attributes to open the attribute dialog box. From the Related records section click within the Layers field.
3. Add one or more layers to the area using the same procedure used to add layers to a borehole. Each borehole inside the area that contains the selected layer will be used to create the solid.

Layers are assigned to geology areas via the Geology ▶ Layer manager command as follows:

1. Choose the Geology ▶ Layer manager and select an area on the screen.
2. Use the navigation buttons at the bottom of the dialog box to select one of the geology layers.
3. From the Related records section click in the Areas field to open the grid dialog box.
4. Add areas containing the selected layer using the same procedure as used to add boreholes to a layer.

The final step is to generate a solid within the volume occupied by the geological layer. For this purpose, the user should:

1. Define the triangulation parameters by choosing Geology ▶ Solids TIN parameters (Figure 2.52).
2. Use the Precision and MaxArea fields in the same manner as for terrain triangulation.
3. Check the LimitedWithTerrain field to use the DTM as the upper limit for the upper surface of any solid.

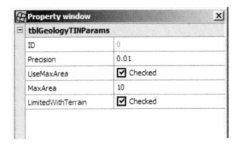

Figure 2.52 Parameters for the triangulation of geology data

Source: The authors

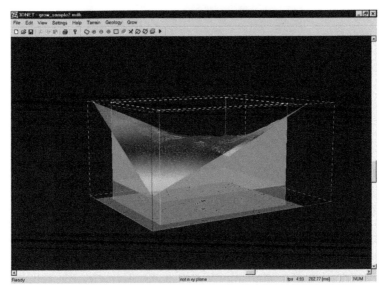

Figure 2.53 An example of two simple geological solids created from geology layers in four
boreholes positioned at four corners of a rectangular area (see also colour plate 25)

Source: The authors

- Create solids by running Geology ▶ Solids Triangulate.
- The solid name is composed of the layer and the area name, for example, Aquifer grow.

An example of two simple solids is presented in Figure 2.53. The solids are created from elevations defined in four boreholes positioned at four corners of a rectangular area. The surface of the upper solid is limited by the terrain elevation.

2.7.5 *GROW* component

This component encompasses all subsurface water modelling tasks, including specification of the urban networks which interact with the urban aquifer. The *GROW* component

Figure 2.54 Grow menu commands

Source: The authors

menu is shown in Figure 2.54. The user can:

- define the hydraulic characteristics of hydrogeological layers,
- define domain boundaries and boundary conditions for the simulation of urban groundwater dynamics,
- simulate leakage through unsaturated soil due to precipitation (the *UNSAT* model),
- generate the finite-element mesh,
- determine leakage from the urban water networks into each finite element,
- simulate groundwater flow and view the simulation results (the *GROW* model), and
- calculate the surface runoff water balance (the *RUNOFF* model)

Before performing aquifer simulations, you have to specify the following model elements:

- model geometry, including the hydrogeological layers assigned either to the 'topsoil' or the aquifer, boundary lines and various point objects such as simple points, wells, sources or observation wells,
- soil characteristics (hydraulic conductivity, effective porosity, etc.),
- boundary conditions,
- the finite-element mesh.

Defining the model geometry

To assign an existing geological layer, that is, an existing solid, to *GROW* as 'topsoil' or an aquifer, the user should:

1. Select Grow ▶ Topsoil layer or Grow ▶ Aquifer layer.
2. From the SolidID dropdown list select the name of the corresponding geological solid. Selecting the solid specifies only the geometry of the layer. The procedure for assigning topsoil or aquifer characteristics to the selected solid, is described below.

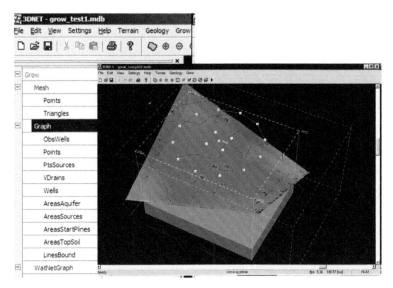

Figure 2.55 **SceneGraph** window showing all available objects under the **Grow-Graph** node and a model window showing an example of a boundary line for defining the modelling domain (see also colour plate 26)

Source: The authors

To add or edit specific point, line or area objects the user should:

1. Select the appropriate object type. For this purpose, in the SceneGraph window, the appropriate sub-node of the Grow-Graph node (for example, the LinesBound node for boundary lines) should be selected. Figure 2.55 shows all object types available.
2. Follow the general procedures for adding or editing graphical objects as explained in Section 2.7.2.

The external model boundary is composed of boundary lines which *must* be created from points by selecting them in a counter-clockwise direction. After completion, the model boundary becomes a positively-oriented closed polyline. Figure 2.55 shows the geometry of a simple model with a circular boundary and a single well in the centre of the model domain. The diameter of the domain is 100 m.

Defining topsoil and aquifer characteristics

There are two types of hydraulic property that must be specified in the *GROW* component of *UGROW*:

*   aquifer parameters, which include the properties of **saturated** groundwater flow such as hydraulic conductivity, effective porosity, specific yield and so on, and
*   topsoil parameters, which include the properties of **unsaturated** flow in the vadose zone near the ground surface, such as saturated hydraulic conductivity, van-Genuchten parameters, maximum water content and residual water content.

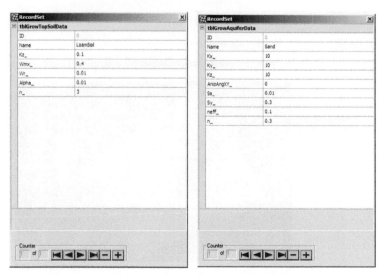

Figure 2.56 Grow ▶ Topsoil (soil) types and Grow ▶ Aquifer (soil) types dialog boxes
Source: The authors

A library of predefined aquifer and topsoil types (e.g. gravel, sand and clay) with corresponding properties is already present in the database. This library should be changed and/or amended to include the soil and aquifer types which represent the conditions of the specific case to be simulated. To view, change and add soil types, the user should select Grow ▶ Topsoil (soil) types or Grow ▶ Aquifer (soil) types and enter appropriate values in the fields of dialog boxes as shown in Figure 2.56. New soil and aquifer types are added by pressing the + button at the bottom of the dialog box.

The complete groundwater simulation domain consists of subdomains with each subdomain having its own set of aquifer properties. The same applies to the topsoil, whose subdomains are usually related to land use. In many practical applications, one aquifer or topsoil subdomain will be much larger than the others. In such cases, it is convenient to specify the 'type' of this subdomain as the default type for the whole domain, and then modify it in those subdomains where necessary. To assign default properties to the *GROW* aquifer layer or *GROW* topsoil layer and select the default attributes, the user should:

1. Press the A button from the SceneGraph window associated with AreasAquifer or AreasTopsoil. The attribute dialog box for the default zero ID object is opened.
2. Expand the Related records section and select AquiferData (TopSoilData) to open a grid dialog box (Figure 2.57).
3. From the drop-down list in the AquiferDataID (TopSoilDataID) field choose the appropriate soil type from the existing library.

To define regions with non-default characteristics, the user should add areas Grow-Graph-AreasAquifer (Grow-Graph-AreasTopsoil) by following the procedure explained in 'Adding lines and areas' in Section 2.7.2. An appropriate soil type is then

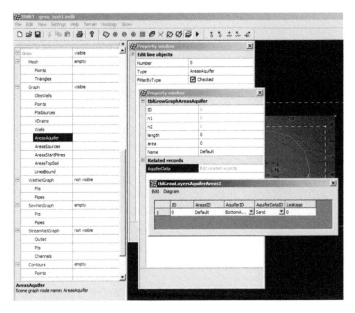

Figure 2.57 Assigning 'types' to the topsoil solid and aquifer solid (see also colour plate 27)

Source: The authors

assigned to the area by editing and repeating the procedure explained above, in other words, by expanding the Related records section, selecting AquiferData (TopSoilData) to invoke the grid dialog box, **and** choosing the soil/aquifer type from the library.

Defining boundary conditions

Boundary lines enclose the simulation domain; the procedure for specifying their geometry is described above. This section concerns the specification of the types and values for boundary conditions which the aquifer simulation model *GROW* will apply along the boundaries. To assign boundary conditions the user should:

1. Select the Grow-Graph-LinesBound node in the SceneGraph window.
2. Press the Edit Lines tool button to open the Edit Line Objects dialog box, select LinesBound in the Type field, and select a boundary line on the screen.
3. Right-click and select Attributes, expand the Related records section and click on the Aquifer field to invoke the grid dialog box. Appropriate data for the boundary can then be entered.

The default boundary condition for all boundaries is a zero flux or no-flow boundary. In other words, if data are not specified for a boundary, that boundary is assumed to be impervious.

The procedure for assigning the boundary conditions to point boundaries (e.g. wells) is analogous to the procedure for boundary lines. It starts with selecting the appropriate node in the SceneGraph window and clicking the Edit Points tool button, and is followed by steps identical to those used for boundary lines.

Modelling vertical water movement in the unsaturated zone (UNSAT model)

The *UNSAT* model simulates the migration of water through the 'topsoil' by solving the unsaturated flow equations. It also calculates the water balance terms. One of these terms is the release of water from the vadose zone into the aquifer, which is subsequently used as input data for the groundwater simulation model *GROW*.

The procedures for entering input data for the *UNSAT* model are as follows:

- Meteorological data (precipitation and potential evaporation over time) are entered through a dialog box displayed by invoking the Grow ▶ Meteorological data command.
- Default values for the soil characteristics are specified by selecting **A** from the Grow-Graph-AreasTopsoil node in the SceneGraph window, while values for non-default regions are entered by editing AreasTopsoil.
- Simulation parameters (initial soil saturation, surface runoff coefficient and soil depth, (usually up to 2 m) are prescribed using the same procedure as for defining soil characteristics.

After defining the input data, an *UNSAT* simulation can be initiated using the Grow ▶ *UNSAT* simulation command and pressing the Action button. At the end of the simulation, a diagram appears on the screen, showing the basic components of the vertical water balance: precipitation, leakage, runoff and actual evaporation from the soil (Figure 2.58). Calculated data representing the release of water from the vadose zone to the aquifer are stored in the database and are ready to be assigned to finite elements during the aquifer simulation.

The topsoil water balance results for a specific region can be visualized by editing the Grow-Graph-AreasTopsoil object and opening the attributes dialog box. After expanding the Related records section and selecting the UnsatWaterBalance item, the grid dialog box with the topsoil water balance results is displayed on the screen.

Modelling interactions with the urban water networks

Three types of urban water network can be created: water supply networks (*WatNet*), sewer networks (*SewNet*) and stream networks (*StreamNet*). Elements of these networks are associated with the three corresponding branch nodes of the *GROW* component (Figure 2.59).

Adding and editing elements of the networks is identical to handling any other graphical object, as explained in Section 2.7.2.

After defining the geometry of the networks, leakage parameters must be set for the line objects. To set leakage rates for pipes of the water supply network, the user should:

1. Select the Grow-WatNetGraph-Pipes node from the SceneGraph window.
2. Press the Edit Lines tool button to open the Edit Line Objects dialog box, and select a pipe on the screen.
3. Right-click to open the Attributes dialog box and enter a value in the Leakage field.

The physical meaning of the pipe leakage parameter is explained in Section 2.6. It takes the form of either a given leakage rate (volume per unit time) per unit length of

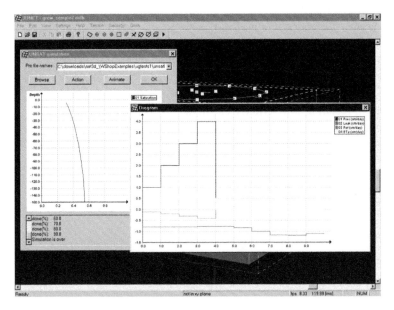

Figure 2.58 **Results of a simulation using the *UNSAT* model (see also colour plate 28)**

Source: The authors

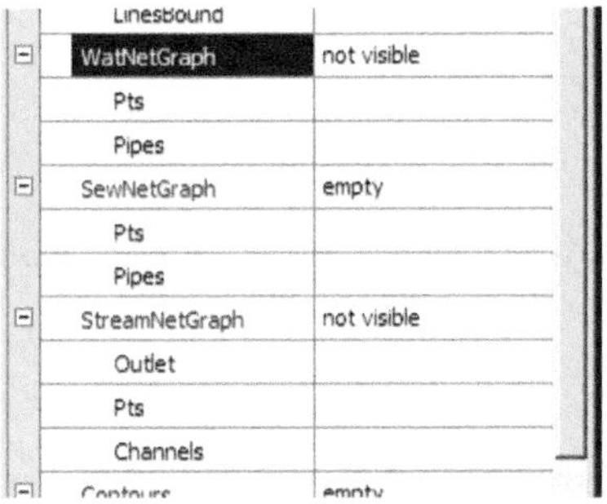

Figure 2.59 Urban water network elements in **SceneGraph**, as branch nodes of the *GROW* node

Source: The authors

the pipe, or the leakage rate per unit pipe length as a function of the head difference between the pipe and the surrounding groundwater. The former method of setting leakage from the pipes is referred to as **Type 2** leakage, while the latter method is termed **Type 3** leakage. For **Type 3** leakage, the leakage rate is taken as constant when the water table is below a specified minimum value. According to the *UGROW* sign convention, the leakage rate has a positive value if water exfiltrates from a pipe, in other words, if it recharges the aquifer.

Generating the finite-element mesh

This is the process of dividing the model domain into smaller subdomains (finite elements) as part of the numerical solution of the partial differential equation describing transient groundwater flow. In *UGROW*, the domain is divided into second-order elements, each with six nodes (Figure 2.60).

Triangulation is carried out using the same mesh-generating algorithm as used for the terrain and geology data. To define the triangulation parameters, the user should:

1. Open the Grow ▶ Mesh TIN parameters dialog (Figure 2.61).
2. Use the Precision and MaxArea fields in the same manner as for terrain triangulation.
3. Optionally, choose to refine elements near wells in the RefineNearWells field.

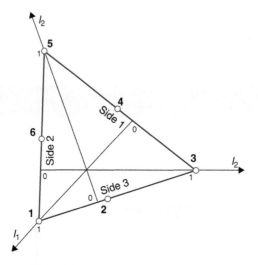

Figure 2.60 Mesh element in the *GROW* model. Bold digits denote node numbers; l_1, l_2, l_3 are the local coordinates

Source: The authors

Property window	☒
⊟ **tblGrowTINParams**	
ID	0
Precision	0.001
UseMaxArea	☑ Checked
MaxArea	10
RefineNearWells	☐ UnChecked
RefineRadius	15
RefineMaxArea	50
IncludeWatNetLeak	☑ Checked
IncludeSewNetLeak	☑ Checked
IncludeStreamNetLeak	☑ Checked
IncludeMeteoLeak	☐ UnChecked

Figure 2.61 Parameters for the triangulation of the finite-element mesh for *GROW*

Source: The authors

4. Check appropriate check boxes to include natural recharge via the topsoil (IncludeMeteoLeak) or leakage from the urban networks (IncludeWatNetLeak, IncludeSewNetLeak, IncludeStreamNetLeak).

Mesh generation is initiated by invoking the Grow ▶ Mesh MESHGEN+UFIND command. It actually consists of two algorithms:

- *MESHGEN* divides the model domain into finite elements (in this case six-point triangles) (Figure 2.62), and
- *UFIND* defines the three-dimensional geometry of each element (area and top and bottom aquifer unit/topsoil elevations), the physical characteristics of the element, and the data related to other urban water objects (the water network, sewage network, etc.) which affect the water balance in that particular finite element.

MESHGEN is the triangulation algorithm implemented over the domain, which is fully enclosed within the model boundary lines. In other words, if we connect the boundary lines together, all points and lines are contained within the geometrical shape formed by the boundary lines. As explained in Section 2.6, a set of points, lines or areas is called a Planar Straight Line Graph (PSLG). Boundary segments enclose the interior of the triangulation domain and clearly separate it from the exterior.

The *UFIND* algorithm finds all elements of the given urban water systems that are relevant to a given finite element. In other words, the algorithm populates the finite-element mesh with the following data:

- The aquifer and aquitard geometry in 3D, which is obtained from the solids defined in the Geology component.
- The region with uniform aquifer and aquitard characteristics (hydraulic conductivity, effective porosity, etc.) to which the element belongs.

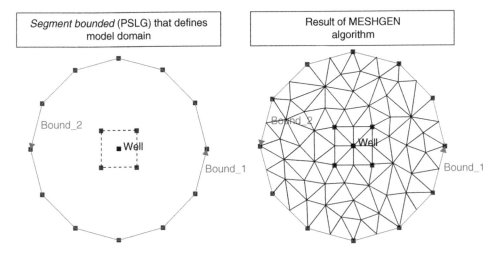

Figure 2.62 Triangulating the model domain (see also colour plate 29)

Source: The authors

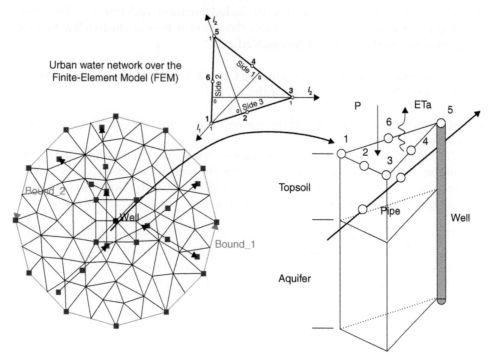

Figure 2.63 **Defining vertical water balance input data for each mesh element (see also colour plate 30)**

Source: The authors

- The relationships between groundwater point and link objects that are creating the PSLG of the modelling domain, and the mesh points and elements: point and line boundary conditions etc.
- The vertical water balance components due to meteorological and soil conditions, and leakage from urban water networks (Figure 2.63).

After mesh generation, the attributes of mesh elements and points can be visualized (Figure 2.64) as follows:

1. Select the **Grow-Mesh-Triangles** node or **Grow-Mesh-Points** node from the **SceneGraph** window.
2. Press the **Edit Lines** or **Edit Points** tool button, and select an element on the screen (to select a triangle, click anywhere within it).
3. Right-click to open the **Attribute** dialog box, and expand the **Related Record** section. Click on the appropriate field (e.g. **AquiferData**) to display the attributes.

Groundwater simulation with the GROW model

Running a simulation and viewing the results requires the following steps:

1. Set the simulation parameters by selecting the **Grow ▶ Simulation parameters** or **Grow ▶ Simulation parameters – advanced** commands.

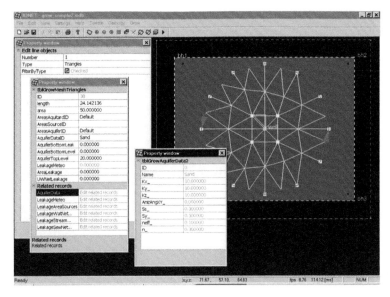

Figure 2.64 **Editing a mesh element and viewing its attributes (see also colour plate 31)**

Source: The authors

2. Set the duration of the simulation and the time-step associated with the pattern functions, which describe the variability of time-dependent parameters, via the Grow ▶ Times command.
3. Perform a steady-state simulation with Grow ▶ Simulate Steady *GROW*, or a transient simulation with Grow ▶ Simulate Unsteady *GROW*.
4. Start an animation of the simulation results by clicking the Play button on the toolbar. ▶
5. View and export simulation results for points or elements of the mesh, or for each object that interacts with the aquifer (pipes or streams associated with the urban water networks).

In the parameters dialog box, the user can enter the maximum error, maximum number of iterations and the collocation parameter. Advanced parameters (the Grow ▶ Simulation parameters – advanced command) define the treatment of leakage originating from pipes and streams of water distribution, sewage and stream networks. Parameters defining leakage type have the value 2 or 3 for head-independent and head-dependent leakage, respectively. For example, the default value for water supply network leakage (the LeakageTypeWat parameter) is 2, because leakage from pressurized distribution networks does not typically depend on the groundwater head. The value of 3 (Type 3) is, by default, associated with sewage and stream networks.

After the simulation parameters are defined and the simulation time is set, the *GROW* simulation engine can be started.

Figure 2.65 shows the results of a simulation carried out using *GROW* in an unsteady-state. The example contains an abstraction well in the centre of the modelling domain and

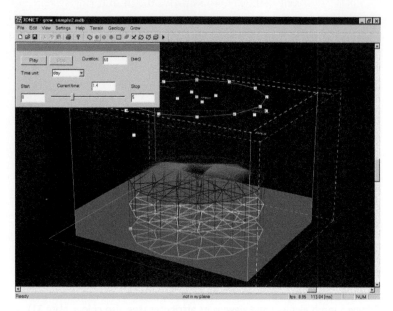

Figure 2.65 Simulation of groundwater flow affected by leakage from a water supply pipe and discharge from a well in the centre of the modelling domain (see also colour plate 32)

Source: The authors

a leaking water supply pipe. The simulation results illustrate groundwater interaction with the pipe. The animation shows rising water table levels due to leakage, which is significantly greater than discharge from the well.

Besides animation, there are several ways to view simulation results for specific objects (boundary lines, wells, pipes, streams, etc.), mesh points or mesh elements. In each case, the process is carried out through the Attribute dialog box. The user has to select (edit) a specific object, open the Attribute dialog box, expand the Related Record section and invoke Simulation Results.

Figure 2.66 shows the simulation results for a selected object of the type: Grow-Mesh-Points. Simulation results showing the leakage (Recharge) from a selected water distribution pipe can be inspected in an identical way.

In addition to viewing the numerical results for specific objects, global water balance results can be viewed by invoking the Grow ▶ Simulation results (water balance) grid dialog box. Global water balance results contain the aquifer recharge summed for all objects of the same type during each simulation time-step. Figure 2.67 shows the global water balance for the same simple example provided in Figure 2.65. From the water balance, it is clear that pipe leakage is excessive, and unrealistically large amounts of water entering the aquifer result in a rapid rise in groundwater level. Water leaves the modelling domain via the boundaries and through the well in the centre, which has a capacity of 20 l/s.

Creating pathlines

The line traced by a fluid particle as it moves is called a pathline. Determining pathlines and travel times along them is the first step in solving advective transport in groundwater.

	ID	PointID	Time	Head
1	98214	215	0	38.1945
2	99439	215	1	44.5487
3	100664	215	2	43.1732
4	101889	215	3	43.4703
5	103114	215	4	43.4106
6	104339	215	5	43.4101
7	105564	215	6	43.4101
8	106789	215	7	43.4101
9	108014	215	8	43.4101
10	109239	215	9	43.4101
11	110464	215	10	43.4101
12	111689	215	11	43.4101
13	112914	215	12	43.4101
14	114139	215	13	43.4101
15	115364	215	14	43.4101
16	116589	215	15	43.4101
17	117814	215	16	43.4101
18	119039	215	17	43.4101
19	120264	215	18	43.4101
20	121489	215	19	43.4101

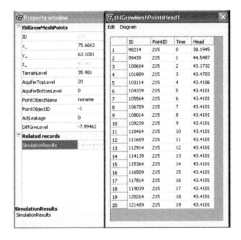

Figure 2.66 Example of a simulation of groundwater head at a selected mesh point

Source: The authors

	ID	Time	LinesBound	PtsSources	Wells	AreasSources	Meteo	WatNet	SewNet	StreamNet	Storage	TotalIn	TotalOut	Error
1	0	1	-0.09328	-0.02	0	0	0	0.1156	0	0	0.002741	0.1156	-0.1133	0.00383
2	1	2	-0.0963	-0.02	0	0	0	0.1156	0	0	-0.0005989	0.1156	-0.1163	0.00103
3	2	3	-0.09585	-0.02	0	0	0	0.1156	0	0	0.0001313	0.1156	-0.1159	0.00352
4	3	4	-0.09614	-0.02	0	0	0	0.1156	0	0	-2.523e-005	0.1156	-0.1161	0.00463
5	4	5	-0.09584	-0.02	0	0	0	0.1156	0	0	6.545e-008	0.1156	-0.1158	0.00227
6	5	6	-0.09582	-0.02	0	0	0	0.1156	0	0	9.532e-009	0.1156	-0.1158	0.00206
7	6	7	-0.09582	-0.02	0	0	0	0.1156	0	0	7.191e-009	0.1156	-0.1158	0.00204
8	7	8	-0.09582	-0.02	0	0	0	0.1156	0	0	0	0.1156	-0.1158	0.00205
9	8	9	-0.09582	-0.02	0	0	0	0.1156	0	0	0	0.1156	-0.1158	0.00205
10	9	10	-0.09582	-0.02	0	0	0	0.1156	0	0	0	0.1156	-0.1158	0.00205
11	10	11	-0.09582	-0.02	0	0	0	0.1156	0	0	0	0.1156	-0.1158	0.00205
12	11	12	-0.09582	-0.02	0	0	0	0.1156	0	0	0	0.1156	-0.1158	0.00205
13	12	13	-0.09582	-0.02	0	0	0	0.1156	0	0	0	0.1156	-0.1158	0.00205
14	13	14	-0.09582	-0.02	0	0	0	0.1156	0	0	0	0.1156	-0.1158	0.00205
15	14	15	-0.09582	-0.02	0	0	0	0.1156	0	0	0	0.1156	-0.1158	0.00205
16	15	16	-0.09582	-0.02	0	0	0	0.1156	0	0	0	0.1156	-0.1158	0.00205
17	16	17	-0.09582	-0.02	0	0	0	0.1156	0	0	0	0.1156	-0.1158	0.00205
18	17	18	-0.09582	-0.02	0	0	0	0.1156	0	0	0	0.1156	-0.1158	0.00205
19	18	19	-0.09582	-0.02	0	0	0	0.1156	0	0	0	0.1156	-0.1158	0.00205
20	19	20	-0.09582	-0.02	0	0	0	0.1156	0	0	0	0.1156	-0.1158	0.00205

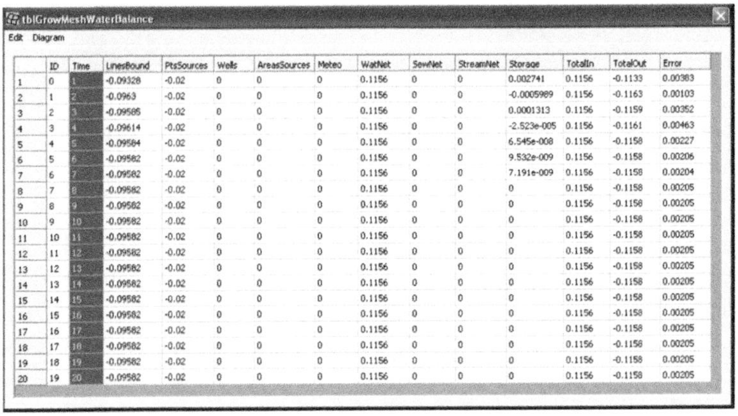

Figure 2.67 Global water balance results

Source: The authors

To create pathlines, the user has to define starting points (Grow-Pathlines-StartPoints) by using the procedure for adding points explained above. Pathlines can be generated only after the simulation has been carried out, because their calculation requires that the fluid velocity field is known.

Since the modelling domain is subdivided into triangles, the groundwater head function, H, for each element can be approximated as a plane defined by the three vertices of the triangle. The pathlines and corresponding travel times are calculated in the direction of steepest descent along an element by the numerical integration of fluid velocity over time:

$$\Delta s = \int_{t}^{t+\Delta t} |\vec{v}| \cdot dt \tag{2.7.1}$$

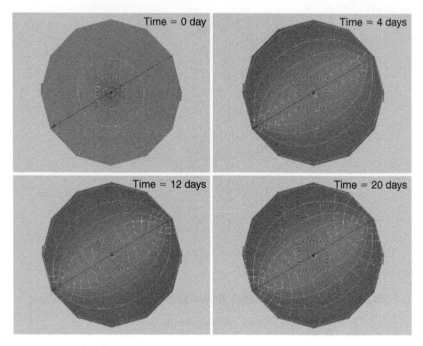

Figure 2.68 Results of implementing the pathline algorithm (see also colour plate 33)

Source: The authors

The time-step Δt used in the numerical integration is adaptive and takes appropriate care of the moment in time that the pathline leaves one mesh element and enters the adjacent element.

Figure 2.68 shows the pathlines and travel times for the example presented in this section.

2.8 MODEL APPLICATION

The data required to build a site-specific simulation model using *UGROW* are listed in Section 2.6. With respect to their function in the model, they can be classified into the following groups:

- Model geometry: points (e.g. that define the terrain, boundaries of geological layers, model boundaries, pipes, etc.), lines, areas, pipe diameters and lengths, stream widths and lengths.
- Boundary conditions and initial conditions.
- Parameters related to: (i) Groundwater flow: hydraulic conductivity, specific storativity, effective porosity/specific yield as related to the fluctuating water table, effective porosity in the context of pore velocity; (ii) Moisture migration through the vadose zone: saturated hydraulic conductivity in the vertical direction, maximum and residual water content, van Genuchten soil parameters; and (iii) Pipes and streams: leakage rates or leakage coefficients.

Data gathering and quality control are always challenging tasks. Even practical problems of moderate size require a large amount of data, which has to be retrieved from diverse sources. Moreover, data from all three groups inherit some degree of uncertainty. This is especially true of data from the latter two groups since, in most cases, they are not measurable quantities and must be obtained by model calibration. This section provides an overview of the calibration methods commonly used in engineering practice (2.8.1), highlights the underlying uncertainty (2.8.2), and shows how sensitivity analysis can be used to evaluate the effect of an error in a parameter. As discussed above, these issues of parameter uncertainty are not unique to *UGROW*, but are common to all groundwater simulation models.

2.8.1 Calibration

Model calibration is the procedure of demonstrating that the model can reproduce field-measured quantities, such as hydraulic heads and flows. It is carried out by finding a set of parameters and boundary conditions that produce simulation results in good agreement with the field-measured data. This is commonly known as solving an inverse problem, as opposed to a forward problem, where the parameters and boundary conditions are fully known. While the latter involves a single model run, the former requires some kind of optimization procedure.

Model calibration can be performed for steady-state or transient flow conditions. It is common practice to use a two-step calibration method where aquifer transmissivity values are evaluated first using steady-state data, while storativity is calibrated using transient data.

For steady-state calibration, the period representing the steady state must be selected very carefully – it could be monthly, seasonal or an annual average. For groundwater flow simulation using *GROW*, the main criterion for selecting the appropriate time interval is flow dynamics: the net effect of transient variations around the steady-state average must be negligible for the calibration to make sense. The selected time interval should not influence the model parameters: they should be valid for any steady-state simulation and also for any transient simulation, provided that the specific storativity/effective porosity values are good. When it comes to calibrating the parameters of *UNSAT* and *RUNOFF*, it has to be noted that these models are conceptually based around the simulation of individual storm events. If continuous rainfall data are not available, they can be run with daily or even monthly values. However, the parameters will be slightly different. For example, the value of vertical hydraulic conductivity, which gives a good estimate of infiltration and runoff during individual storm events, would need to be modified if the simulation is carried out with daily or monthly rainfall values. Rather than simply representing hydraulic conductivity, this parameter now accounts for the temporal variability of wet and dry periods.

For some groundwater flow regimes, the assumption of steady-state conditions may be inappropriate due to large variations in water level. In such cases, the model may be calibrated to transient conditions. Transient calibration usually starts with the calibrated steady-state solution. Alternatively, simulations may start from an arbitrary initial condition and run for a sufficiently long period, prior to calibration, for the influence of the initial conditions to diminish.

There are two main methods of performing a model calibration: manual parameter tuning by trial-and-error, and automated parameter estimation. The manual method was the first to be developed and is still preferred by many practising modellers. This method was used in all the case studies presented in Chapter 3. The drawback of this method is that it does not force the modeller to adhere to any protocol. The procedure is often poorly documented and the quality of the result depends heavily on the modeller's experience and intuition. Automated estimation is much quicker and less tedious. However, since resolution of the inverse problem does not have a unique solution, the algorithm may produce a solution which formally satisfies optimization criteria but does not reflect the physics of the system. For this reason, it is best to combine the two methods, in other words, use automated procedures combined with modelling experience and intuition.

2.8.2 Uncertainty

All *UGROW* simulation models are deterministic; in other words, a single input of parameters and boundary conditions produces a single set of values for groundwater heads and velocities. However, model parameters and boundary conditions are notoriously uncertain quantities. An alternative approach, known as probabilistic or stochastic modelling, uses probability distribution for model parameters and boundary conditions. Consequently, a simulation result is obtained as probability distribution, rather than a single value for each flow variable.

It is possible to perform probabilistic modelling using a deterministic code. The most popular method for doing this is the Monte Carlo simulation. The probability distributions of model parameters and boundary conditions (e.g. transmissivity, storativity, recharge) are assumed and then randomly sampled for a large number of model runs (1,000 or more), each run now producing a single realization of each random variable. The large number of realizations enables calculation of the probability distribution for each model output variable.

Although the Monte Carlo procedure itself is very simple, it requires significant computer resources. For this reason it is not often used in engineering practice. More efficient stochastic models exist, but these are also more sophisticated and require rather high levels of mathematical skill. Another reason for the lack of popularity of stochastic methods is that decision-makers usually prefer single numbers to probability distributions. Practising modellers similarly prefer to exercise judgement, rather than to manipulate parameters of probability density functions.

Deterministic modelling is probably going to remain more popular than stochastic modelling in engineering practice. However, it should not be forgotten that uncertainty is inherent in model parameters and boundary conditions, as well as in the model output.

2.8.3 Sensitivity

As discussed above, the parameters and boundary conditions of a deterministic model always contain uncertainty. In other words, parameter values obtained by model calibration include an error. In practical situations, the exact value of a parameter is not known, so it is impossible to calculate the error. However, it is possible to evaluate how significant an effect this error has on the model results. This is achieved through sensitivity analysis.

The starting point for sensitivity analysis is the set of parameters obtained from model calibration. This set of parameters is used to produce the reference model output. A number of parameters is then selected for the sensitivity analysis. Each parameter is varied, one at a time, to produce a corresponding output. Relatively large variations in model output indicate that the model is very sensitive to that particular parameter.

If the model is highly sensitive to a certain parameter, it is possible to evaluate its value very precisely by calibration. If the value is not accurate, however, this will have a major impact on the quality of model output. As a corollary, models that are relatively insensitive to a parameter will lead to weak, imprecise calibration; however, errors in calibration will not have a significant influence on simulation results.

Sensitivity analysis can be used for the selection of suitable observation points, for providing model calibration data, or to gain insight into model performance.

Chapter 3

UGROW applications – case studies

Leif Wolf[1], Christina Schrage[2], Miloš Stanić[3] and
Dubravka Pokrajac[4]

[1]*Institute for Applied Geosciences, University of Karlsruhe, Karlsruhe, Germany*
[2]*Project Manager Geo Ecology, Karlsruhe, Germany*
[3]*Institute of Hydraulic Engineering, Faculty of Civil Engineering, Belgrade, Serbia*
[4]*School of Engineering, University of Aberdeen, Aberdeen, United Kingdom*

3.1 TESTING AND VALIDATION OF *UGROW* IN RASTATT, GERMANY

3.1.1 Scope and motivation

When developing new software programs, one of the primary objectives is to ensure that the results obtained are comparable to standard solutions accepted within the literature. It is also important to ensure that programs are user-friendly for inexperienced users. To this end, the *UGROW* model was tested and evaluated using data from the city of Rastatt in southwest Germany, for which a calibrated FEFLOW® (groundwater flow model) groundwater flow model already exists as part of an AISUWRS (Assessing and Improving the Sustainability of Urban Water Resources and Systems) project (Wolf et al., 2006a, 2006b). Testing the *UGROW* model involved preparing the input data for the simulation run, model calibration, sensitivity analysis and comparing *UGROW*'s modelling results with those obtained by FEFLOW®. In order to conduct such a comparison, input values for the *UGROW* model were taken from the existing FEFLOW® model wherever possible. To complete the analysis, *UGROW* results were compared with results obtained from the AISUWRS study.

3.1.2 Geographical setting

The city of Rastatt (population approaching 50,000) is located 30 km south of Karlsruhe, close to the eastern border of the Upper Rhine Valley in southwest Germany (Figure 3.1). Rastatt's climate is continental, characterized by hot summers and cool winters. The mean annual temperature is 10°C. Annual rainfall is between 850 and 1000 mm and peaks in the summer months (Eiswirth, 2002). Precipitation in the Rastatt region is locally variable due to the bordering slopes of the Black Forest.

Four aquifers have been identified in the Rastatt area: the Upper Gravel Layer (OKL), the Middle Gravel Layer (MKL), the Older Quaternary (qA) and the Pliocene aquifer. This aquifer sequence is overlain by Holocene cover sediments of variable grain size and thickness. An upper interlayer comprising fine-grained sediments separates the OKL from the MKL. A conceptual hydrogeological model derived from existing geological information and knowledge of major water flows within the area is shown in Figure 3.2.

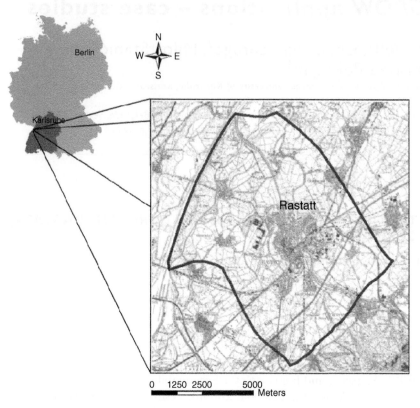

Figure 3.1 **Geographical setting**

Source: Klinger and Wolf, 2004

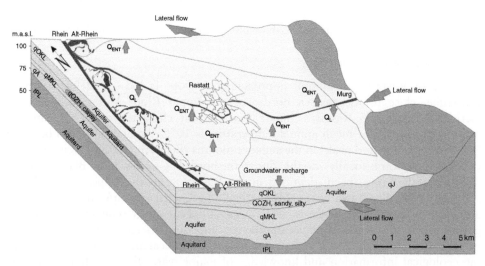

Figure 3.2 **Hydrogeological setting in Rastatt**

Source: Modified after Eiswirth, 2002

3.1.3 Existing investigations and models available for benchmarking

Three previous research projects have been conducted for the city of Rastatt. These have focused on:

- the interaction between sewer systems and groundwater and
- the development of a holistic water balance for Rastatt.

The first of these studies was the European Union (EU)-supported AISUWRS project (see Section 1.4.2), which attempted to bridge the gap in knowledge and understanding between groundwater resources and surface/near-surface urban water networks (Eiswirth, 2002; Eiswirth et al., 2004; Wolf et al., 2006b). Field investigations were conducted in four cities located in Europe and Australia. The objectives were to measure and describe the impact of city water infrastructure on urban groundwater resources, and to develop a suite of models to describe and link the urban groundwater system with the unsaturated zone and urban water supply and sewer networks (Burn et al., 2006).

The uppermost model represented in the AISUWRS system is the Urban Volume Quality model (UVQ) developed by CSIRO, Australia (Mitchell and Diaper, 2005; Diaper and Mitchell, 2006). Its main input parameters are climate records, water consumption characteristics (e.g. water use for laundry and typical contaminant loads through toilets) and urban surface permeability coefficients. The AISUWRS model calculates water flows and contaminant loads through urban wastewater and stormwater systems, and assesses the direct contribution to groundwater (e.g. recharge).

The information obtained from the UVQ is fed into the specially developed Network Exfiltration and Infiltration Model (NEIMO), which estimates the amount of wastewater exfiltration from, or groundwater infiltration into, sewers (DeSilva et al., 2005). Leakage rates are based on the distribution of pipe defects observed by closed circuit television (CCTV) investigations or, where no CCTV data are available, the application of characteristic curves taking into account pipe material and age. Typically, exfiltration rates from damaged sewers are found to be between 0.139 l/d/m and 3.64 l/d/m (Wolf et al., 2006b).

The output from NEIMO is forwarded to purpose-designed unsaturated zone models (SLeakI, POSI and UL_FLOW) that calculate water flows and solute travel times to the water table from both point sources (leaking sewers) and distributed sources (rainfall). These models incorporate the combined effects of sorption and decay of contaminants in the calculations as the water travels through the unsaturated zone. Finally, numerical groundwater flow and transport models are employed to determine the movement of contaminants within the aquifer.

The AISUWRS concept was applied to all four case study cities. Major water fluxes were quantified together with loadings of marker substances (e.g. chloride, potassium, boron, sulfate and zinc). Extensive groundwater sampling field studies were conducted to confirm the results of the predictive modelling exercises, with the sampling being undertaken at specifically constructed test sites and groundwater monitoring networks. In the modelling exercises, various water management scenarios were modelled. These included the effects of decentralized rainwater infiltration and sewer rehabilitation and water balance changes due to climate change (Rueedi et al., 2005; Cook et al., 2006; Klinger et al., 2006; Morris et al., 2006; Souvent et al., 2006).

In an earlier project phase, the sub-catchment of Rastatt-Danziger Strasse served as a demonstration example for the AISUWRS models (Klinger and Wolf, 2004;

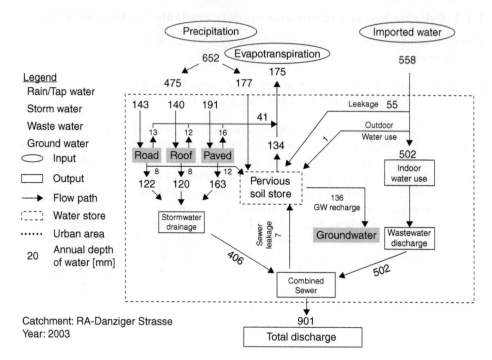

Figure 3.3 Total water balance calculated for the densely populated Rastatt-Danziger Strasse catchment using the UVQ model

Source: Wolf et al., 2006c

Wolf et al., 2006b, 2006c, 2006d) and the upscaling of sewer leakage quantifications (Wolf and Hötzl, 2006). It is this sub-catchment that was used to test and evaluate *UGROW*. The Rastatt-Danziger Strasse catchment is drained by a combined sewer system, covering an area of 22.4 hectares and comprising a mixed settlement of both residential and commercial buildings. The catchment can be subdivided into six 'neighbourhoods', each with a relatively homogenous housing structure. The water and solute balance was calculated using the UVQ model (Figure 3.3), based on available climate data, demographic data, drinking water consumption and surface sealing (impermeability) maps. Land-use data are entered with options for public open space, paved areas, garden areas and roof areas. Runoff calculations take the soil moisture into account and may be performed either with a partial area or a two-layer soil moisture storage approach.

Early field studies demonstrated a significant impact on aquifer water quality by leaking sewer systems (Wolf et al., 2004; Morris et al., 2006). For example, various pharmaceutical residues were detected in both seepage water and urban groundwaters (Cook et al., 2006; Wolf, 2006a) with iodated X-ray contrast media proving to be especially useful marker substances. Microbiological investigations from the case study sites also showed the widespread occurrence of faecal indicators. A test site was established in Rastatt (the Rastatt-Kehler Strasse site) that offered, for the first time, long-term monitoring of the quantity and quality of sewage leaking from an *in situ* public sewer under operating conditions.

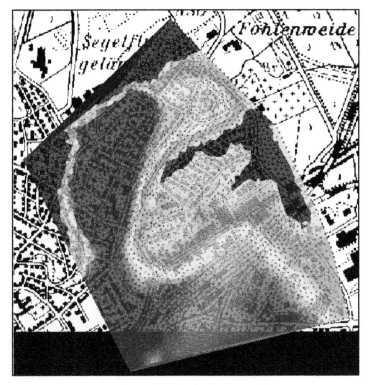

Figure 3.4 *UGROW* base map and digital terrain model

Source: The authors

3.1.4 *UGROW* model setup

UGROW model area and surface elevation

Testing of the *UGROW* model took place in the 2 km² 'Danziger Strasse' catchment located in the north-eastern part of Rastatt. A calibrated FEFLOW® previously developed by the University of Karlsruhe exists for this area, and was developed using data from field studies and models from local water works.

The first step in creating the *UGROW* model was to determine the model boundaries and assign surface elevation data. This was undertaken with the *TERRAIN* component of the *UGROW* model system. A base map was loaded in the form of a .jpeg file and positioned by defining the bottom-left and top-right coordinates. Elevation points were exported from the ground surface (terrain surface) of the FEFLOW® model, which, had previously been interpolated from the known elevations of sewer manhole covers. The elevation points were added to the MS-Access table which underlies the *3DNet* application. Mesh triangulation was carried out using a maximum mesh area of 20 m².

Using the 'Mesh Properties' criteria dialog box, map colour properties were assigned to display the resulting digital elevation model. The colours were then chosen to conform with the FEFLOW® display. Figure 3.4 shows the *UGROW* terrain model in plan view with the underlying base map. Figure 3.5 shows the same view in the FEFLOW® model, underlain by the sewage network.

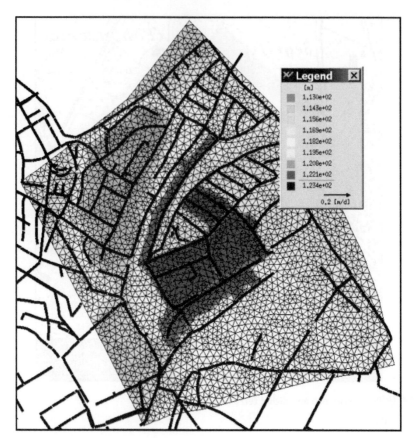

Figure 3.5 FEFLOW® model and sewage network

Source: The authors

Representation of the aquifer system

Underlying the Danziger Strasse study area are two aquifers separated by a silt layer of slightly lower hydraulic conductivity. The existing FEFLOW® model includes five layers, mainly to enable a fine vertical discretization for future numerical modelling of water quality. At the time of testing, both *UGROW* and *3DNet* were limited by an inability to represent more than one aquifer. Thus, the hydrogeology component had to be simplified by combining the two aquifers into a single homogeneous aquifer unit (Figure 3.6). This approach was justified since the hydraulic separation between the upper and lower aquifers is relatively weak in the Danziger Strasse region, and neglecting the discontinuous silt layer has only a minor influence on flow modelling results. This simplification would be less appropriate for the purposes of simulating and predicting contaminant transport.

Within the geology component of *UGROW*, eighteen boreholes were used to define the top and bottom elevations of the aquifer. These included six existing boreholes in the south-eastern part of the model domain and six fictive boreholes located at the model corners and along the model border. Elevation data for the fictive boreholes were

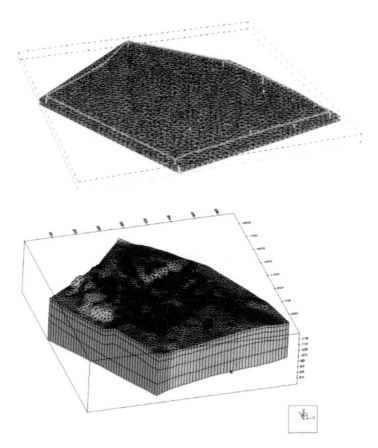

Figure 3.6 Hydrogeological conceptualization for a single aquifer in *UGROW* (above) and a multi-
layered aquifer system (below)

Source: The authors

taken from the stratigraphy at respective positions in the FEFLOW® model, which,
were based on the regional interpretation of drilling logs. Six additional fictive boreholes were established outside the modelled domain to ensure that the geology solid
extended beyond the domain. Elevation data for these outside boreholes were identical
to the corresponding boreholes on the model border. The area for solid creation was
defined using the outermost boreholes. Solid creation is vertically limited by the ground
surface (the terrain surface) and uses a maximum interpolation area of 1,000 m².

For ease of comparison, the aquifer characteristics were assigned the same values that
were used in the calibrated FEFLOW® model. Vertical and horizontal hydraulic conductivity values for the sand aquifer were set to 29.52 m/d and 147.50 m/d, respectively.
The specific yield and effective porosity were assumed to be 20%. In *UGROW*, a link
has to be established between the aquifer layer and the previously created geology solid.

Along the boundary of the model domain, boundary conditions were defined in
accordance with the existing FEFLOW® model. Their locations and characteristics are
depicted in Figure 3.7. For the eastern and western boundaries, no flow conditions

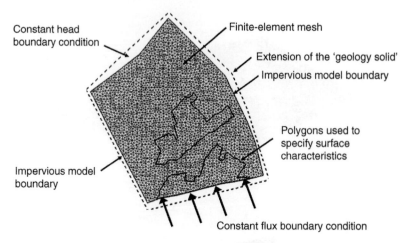

Constant head
boundary condition

Finite-element mesh

Extension of the 'geology solid'

Impervious model boundary

Polygons used to
specify surface
characteristics

Impervious model
boundary

Constant flux boundary condition

Figure 3.7 Boundary conditions, extension of the geology solid and polygons specifying surface
characteristics

Source: The authors

were assumed, while the northern (downstream) border was represented by a constant head boundary of 111.7 m. The southern (upstream) border was established as a constant flux boundary with a flow of approximately $6 \times 10^{-5} \text{m}^3/\text{s/m}$. This flux was obtained by dividing the total flux across the southern FEFLOW® model border ($6,315 \text{ m}^3/\text{d}$) by the length of the border line (1,213 m). Since boundary conditions were considered to be invariant with time, they were assigned an 'identical' pattern of just one stage with a value of '1'.

Within *UGROW*, the unsaturated zone is described by 'AreaTopSoil'. While it is possible to specify the amount of groundwater recharge directly as a constant 'AreaSource', it is generally recommended that recharge be calculated using meteorological data. Recharge can be calculated using a soil water balance invoked by the simulation model *UNSAT*. This model is based is on the Richards equation and requires Van Genuchten parameters, which were acquired from Carsel and Parish (1988) for the purposes of the study. The *UNSAT* term 'active depth' was interpreted as the depth to the water table, which is usually shallow in the city of Rastatt. Following a sensitivity analysis (see Section 2.8), and calibration with rural lysimeter data, a thickness of 1 m was selected. In general, the vertical extent of the root zone defines the active depth.

The active depth for sands in the unsaturated zone was initially set to 3 m (later reduced to 1 m), and the initial soil saturation was specified as 15%. The maximum water content was set at 43%. Surface-sealing data were obtained from a detailed field survey previously conducted by the city of Rastatt. These data were further processed during the AISUWRS project and are shown in Figure 3.8 (Klinger and Wolf 2004; Wolf et al., 2005). Based on the analysis, a model runoff coefficient of 0.5 was established for most of the model domain. Exceptions were made in the southern half of the study site where two densely populated areas were assigned a higher runoff coefficient of 0.8 (Figure 3.9).

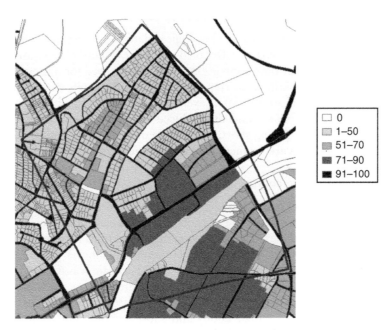

☐	0
☐	1–50
▨	51–70
■	71–90
■	91–100

Figure 3.8 Detailed surface sealing map showing % of sealed surface

Source: Klinger and Wolf, 2004

Figure 3.9 Simplified surface sealing map for the *UGROW* modelling exercise. A runoff coefficient of 0.5 was used in areas shown in white; a value of 0.8 was used for the more densely populated areas shown in solid grey.

Source: The authors

Unsaturated zone characteristics and climate records

UNSAT also requires the input of meteorological data to calculate leakage from the unsaturated zone (aquifer recharge). Detailed records for daily rainfall and evaporation quantities available for the period 2002 to 2004 were used for this exercise. Early calculations employed monthly time steps, but did not adequately describe the runoff processes during short duration rainfall events. Consequently, daily time steps were adopted for *UGROW/UNSAT*, and this reduced the errors significantly. Even with this improvement, runoff during short, heavy rainfall events (e.g. thunderstorms with just one-hour duration, as are frequent in Rastatt) tends to be underestimated.

Water supply and drainage network

Detailed data exist for the public sewer network in the central part of the model. For 169 manholes, triplet information (UTM coordinates and elevation) was imported into the MS-Access database and used for specifying the connections between the individual sewage pipes. The pipes were added as links between the manholes. The total length of the urban drainage network included within the model domain was 6,774 m. The sewer network is shown in Figure 3.10. Monte Carlo simulations were used to extrapolate existing visual inspection data and sewer leakage rates to the sub-catchment of Danziger Strasse (Wolf and Hötzl, 2006) and subsequently, with additional refinements, to the entire City of Rastatt (Wolf, 2006).

In total, 262 sewer leaks were represented in the sub-catchment of Danziger Strasse, and results of the analysis indicated a broad range of possible groundwater recharge rates. The highest probability (23%) was found for a 4.2 mm/a groundwater recharge rate (equivalent to a total leakage of 2.57 m³/d or 0.8% of the typical dry weather flow of 320 m³/h). With a probability of 95%, groundwater recharge will be below 65 mm/a. The estimated maximum possible rate of recharge was 176 mm/a. All sewage exfiltration data were included in *UGROW* model simulations using Type 2 boundaries (as described in Section 2.7).

Digital information on the drinking water supply network was only available in dxf format and did not include all the necessary pipe specifications. A visual comparison of available datasets suggested the existence of a close spatial correlation between the drinking water supply network and the sewer network. Thus, losses from the water supply network were conceptually modelled using virtually the same areal distribution of pipes as contained within the sewer network, but set at a shallower depth. Leakage estimates were based on the number of inhabitants, average daily consumption and water losses measured over the entire city. In the modelling exercise, a leakage rate of 10% of the supplied water volume was assumed (Table 3.1). In total, the model included 6,672 m of pressurized drinking water pipes leaking at a rate of 0.0046 m³ d⁻¹ m⁻¹ (Jaiprasart, 2005).

3.1.5 Model results

Unsaturated zone water balance

Figure 3.11 shows the *UNSAT* simulation results summed monthly over a period of two years. Rates are shown in cm/day where leakage to the water table (recharge) equals precipitation minus surface runoff minus actual evapotranspiration. To follow the sign convention, leakage to the water table is shown as a negative value in the

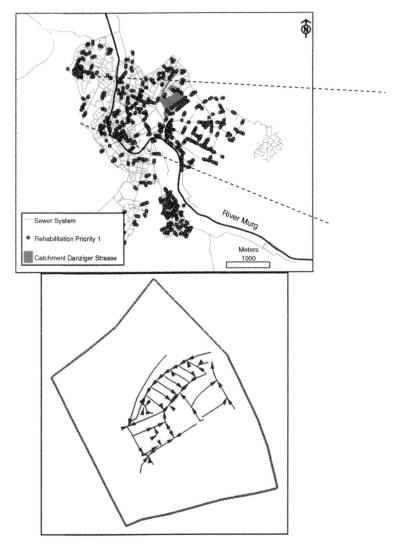

Figure 3.10 Sewer network in Rastatt including major sewer leaks and the modelled part of the sub-catchment of Rastatt-Danziger Strasse

Source: The authors

Table 3.1 Leakage rates determined by Star.energiewerke and the corresponding groundwater recharge averaged over the estimated supply area

Year	Water supplied			Water losses	
	[m³]	[mm/a]	[%]	[m³]	[mm/a]
1996	2579000	152.7	11.32	291943	15.8
1997	2498000	147.7	9.27	231565	12.5
1998	2434000	144.5	7.76	188878	10.2
2000	2475933	147.0	12.84	317910	17.2

Source: Wolf et al., 2005

figure. Leakage (recharge) values are slightly lower for the two areas with higher runoff coefficients (0.8 versus 0.5). Components of the *UNSAT* water balance for 2003 are shown in Table 3.2 (after Jaiprasart, 2005).

Values obtained with *UNSAT* for 2003 (Table 3.2) are in general agreement with the numbers produced for the same period for the UVQ component of AISUWRS (compare Table 3.2 with Figure 3.3). It should be noted that the water balance areas do not match precisely and some differences between the two models should be expected. The

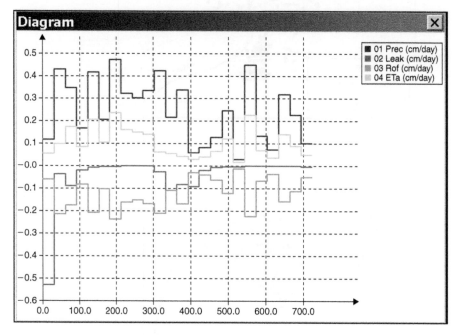

Figure 3.11 Monthly *UNSAT* water balance simulation result. The x-axis is shown in days and the y-axis is shown in cm/day

Source: The authors

Table 3.2 UNSAT water balance for 2003

Components	Water Balance (mm/a)	
	Inflow	Outflow
Precipitation	652.7	
Soil storage	10.7	
Leakage to the aquifer		100.9
Actual evapotranspiration		92.6
Surface runoff		469.9
Totals	663.4	663.4

Source: After Jaiprasart, 2005

unsaturated zone module of AISUWRS includes only the part of the FEFLOW® model domain for which sewer network information is available. In contrast, *UNSAT* performs calculations across the entire *UGROW* area. *UNSAT* predicts a slightly higher amount of surface runoff (470 mm/a versus 406 mm/a), and aquifer recharge (leakage to the saturated zone) is slightly higher with AISUWRS (136 mm/a versus 101 mm/a). The most significant difference was found for actual evapotranspiration, which is much higher in the AISUWRS calculations (175 mm/a versus 93 mm/a).

Sensitivity analysis

The *UNSAT* water balance simulation requires the specification of several unsaturated zone parameters. The initial values for these parameters are summarized in Table 3.3. To provide an estimate of uncertainty in the model results, a sensitivity analysis was performed by varying key input parameters within a broad but realistic range. The analysis was performed for the default AreaTopSoil with a runoff coefficient of 0.5, and for the two areas in the southern part of the model which were characterized by a higher runoff coefficient of 0.8. Since postprocessing tools for *UGROW* are somewhat limited, the sensitivity analysis was evaluated only in terms of the effect on net groundwater recharge (mm/y).

As shown in Figure 3.12, the correlation between recharge rate and runoff coefficient is almost linear, particularly for runoff coefficients below 0.6. The higher the coefficient, the higher the surface runoff, and the less water available for groundwater recharge. Similarly, close to linear correlations are observed between recharge and active soil depth and initial water saturation. Overall, the predicted recharge rate is higher than expected for the dry year of 2003. This is probably due to the selection of initial topsoil parameters (Table 3.3). The starting condition for the simulation is a thick unsaturated zone with high initial saturation storing a large volume of water. In the course of the model run, this water drains from the unsaturated zone and is responsible for the high groundwater recharge. Another problem is the probable minimum depth limit for the unsaturated zone, as the *UNSAT* simulation does not produce results

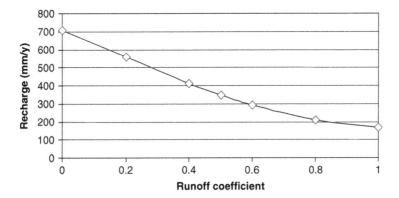

Figure 3.12 Sensitivity of groundwater recharge calculations to the runoff coefficient

Source: The authors

Table 3.3 **Topsoil parameters used as default values for the sensitivity analysis**

Active soil depth	3 m
Initial saturation	15%
Maximum water content	43%
K_z	8.25×10^{-5} m/s

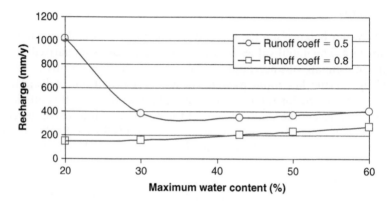

Figure 3.13 Sensitivity of groundwater recharge calculations to maximum water content

Source: The authors

for an active soil depth of only 1 cm. The strong influence of initial topsoil parameters on groundwater recharge implies that high-quality input data must be available to provide a realistic water balance, or a sufficient buffer of computational time must be allocated for model calculations.

The situation is slightly different for the analysis of the sensitivity of the recharge estimate to the maximum water content. In general, recharge gradually increases with higher water storage soil capacities, especially for high runoff coefficients (e.g. 0.8 in Figure 3.13). However, for lower runoff coefficients (e.g. 0.5, where more water is entering the unsaturated soil), a water capacity of less than about 30%, leads to a major reversal of this trend with lower values of water storage capacity leading to unrealistically high values of recharge. The reasons for this behaviour are unclear but may be due to:

- the high initial water content which subsequently drains freely from the soil storage
- reduced evapotranspiration, or
- free drainage as soon as the storage capacity is exceeded.

Further investigations were not conducted due to time constraints. However, it is important that you understands that *UNSAT* in its tested form can produce anomalous results for certain combinations of input parameters. For example, the relationship between the vertical hydraulic conductivity of the soil (K_z) and recharge follows

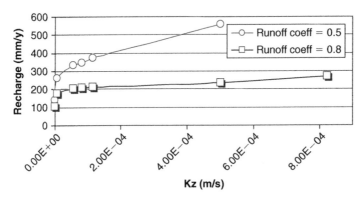

Figure 3.14 Sensitivity of groundwater recharge calculations to vertical hydraulic conductivity of the active soil layer

Source: The authors

Table 3.4 Topsoil parameters for the final model

Active soil depth	1 m
Initial saturation	10%
Maximum water content	43%
K_z	8.25×10^{-5} m/s

hyperbolic curves (Figure 3.14), such that for high hydraulic conductivity values (e.g. 8.25×10^{-4} m/s) *UNSAT* is unable to produce a result for areas where runoff coefficients are relatively low (e.g. 0.5).

Overall, the results of the sensitivity analysis show that several topsoil parameters have a significant influence on calculated groundwater recharge. Lysimeter measurements in the Rastatt area indicate a recharge rate of about 340 mm/y for open spaces (sealing degree of 0.0 from Eiswirth, 2002), and about 90 mm/y of recharge is expected in urban areas. These target values can be obtained by several parameter combinations, in other words, the model setup is not unique and leads to uncertainty. In subsequent analyses, the input parameters listed in Table 3.4 were selected. These provide a recharge of 235 mm/y in the default area (runoff coefficient of 0.5) and roughly 90 mm/y in the two areas with a higher degree of sealing (runoff coefficient of 0.8). Integrated over the entire model domain, recharge amounts to an acceptable 190 mm/y.

Comparison of computed groundwater flow fields

Several simple comparisons between identically set up FEFLOW® and *UGROW* models demonstrate good agreement for both water levels and water balance terms. This indicates that *UGROW* is solving the governing numerical equations for groundwater flow with sufficient accuracy.

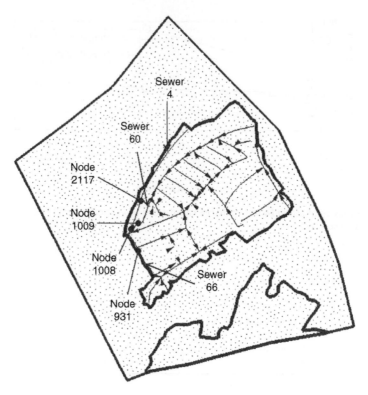

Figure 3.15 Nodes and sewers selected for model validation

Source: The authors

Transient *UGROW* simulations with daily input data for climate parameters produce a seasonally changing water table, and the best way to evaluate the validity of
the model assumptions is to compare results with actual measurements. In the area
modelled by *UGROW*, several groundwater monitoring wells are equipped with automated water level loggers (Wolf, 2004, 2006; Wolf et al., 2005). For comparative purposes, model data were selected for nodes close to the positions of the monitoring
wells (Figure 3.15).

As shown by Figure 3.16, the model reproduces parts of the field data set but does
not provide a perfect match. A general decline in groundwater levels between March
and September 2004 is predicted, but while water levels declined by about 40 cm in
reality, the model predicts only a 20 cm drop. Likewise, the rise in groundwater level
following rainfall events is also predicted but with insufficient magnitude. Possible
reasons for the mismatch are as follows:

- An overall base offset of measured and modelled responses can be explained by the
  base calibration of the steady-state model, which was used to provide parameters
  for the boundary conditions. The steady-state model was calibrated to reproduce

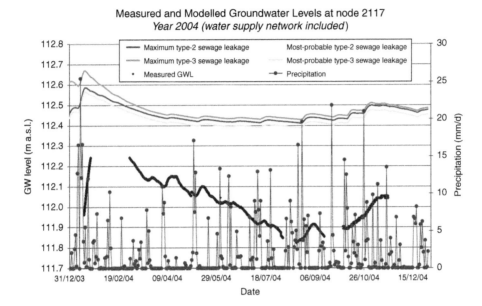

Figure 3.16 Comparison of modelled groundwater levels and measured data (see also colour plate 34)

Source: The authors

the mean hydraulic conditions between 1960 and 1990, but may be an inappropriate choice for providing boundary conditions for the year 2004.

- The upstream model boundary was specified as a time invariant constant flux. In reality, this boundary must vary with time. This could explain the insufficient magnitude of the seasonal trend.
- Several groundwater recharge processes are active in addition to the matrix flow considered by the *UNSAT* module. In urban areas, this can be leakage from sewer systems that support high water levels during storm events. However, it may also include preferential flow along building foundations or other inhomogeneities in the urban subsurface. As shown in Figure 3.16, the different assumptions for sewer leakage in the Rastatt case study model suggest that sewer leakage exerts only a minor influence on the water levels at the focus node 2117.

Most of the above problems can be resolved by taking additional time to revisit the model input parameters, or by extending the model area. Only a very small percentage of the discrepancies observed are actually attributable to the *UGROW* model system.

Scenario simulations

Recognizing the large number of different input parameters and the model's reliance on processes that have a paucity of reliable field data (e.g. lysimeters are rarely available in urban areas), it is suggested that the model simulations be run for a variety of possible boundary conditions (Table 3.5). This may provide future users of the model

Table 3.5 Key water budget values for various urban water supply and sewage network scenarios

Items	Year 2003			Year 2004		
	Inflow	Outflow	Error	Inflow	Outflow	Error
1) No water networks						
Boundaries		312,918			325,201	
Unsaturated zone	321,257			337,137		
Storage		3,307			6,870	
Total	**321,257**	**316,225**	**5,032**	**337,137**	**332,071**	**5,066**
2) Water supply network only						
Boundaries		324,245			336,547	
Unsaturated zone	321,257			337,137		
Water supply network	11,348			11,379		
Storage		3,308			6,869	
Total	**332,605**	**327,553**	**5,053**	**348,517**	**343,416**	**5,101**
3) Type-2 sewerage only						
Boundaries		314,887			327,154	
Unsaturated zone	321,257			337,137		
Sewerage	1,970			1,976		
Storage		3,306			6,871	
Total	**323,227**	**318,193**	**5,034**	**339,113**	**334,025**	**5,088**
4) Water supply network and type-2 sewerage						
Boundaries		326,224			338,526	
Unsaturated zone	321,257			337,137		
Water supply network	11,348			11,379		
Sewerage	1,970			1,976		
Storage		3,308			6,369	
Total	**334,576**	**329,532**	**5,044**	**350,492**	**345,395**	**5,097**
5) Type-3 sewerage only						
Boundaries		315,935			328,177	
Unsaturated zone	321,257			337,137		
Sewerage	1,991			1,973		
Storage		2,304			5,856	
Total	**323,248**	**318,240**	**5,008**	**339,111**	**334,034**	**5,077**
6) Water supply network and type-3 sewerage						
Boundaries		327,197			339,480	
Unsaturated zone	321,257			337,137		
Water supply network	11,348			11,379		
Sewerage	1,952			1,935		
Storage		2,298			5,852	
Total	**334,558**	**329,495**	**5,063**	**350,452**	**345,333**	**5,119**

with additional information regarding the influence these processes may exert on the overall outcome. Caution must be exercised when interpreting Table 3.5, as the results suggest that the quantitative impact of water mains and sewer leakage on the total water budget is small. While this may true for Rastatt, completely different results may be found in other cities.

3.1.6 Summary and conclusions

The sustainable protection of groundwater resources in urban areas requires the integrated management of both the urban water infrastructure and the underlying aquifer. The major strength of the urban water management tool *UGROW* is that it fully integrates a groundwater flow model with models capable of simulating urban runoff characteristics, processes in the unsaturated zone, and flows to and from urban water infrastructure networks. The comparison of *UGROW* with the AISUWRS model suite (Burn et al., 2006; Mitchell and Diaper, 2005; Wolf et al., 2006c) showed acceptable agreement, and model validation with the commercially distributed FEFLOW® simulation software (Diersch, 2005) was successful.

The application testing in Rastatt demonstrated that users who are not involved in development of the model code are able to operate the *UGROW* system successfully, provided that appropriate support is available. As a result of this test application, the user friendliness of the system was significantly improved.

Validation with real world data showed that care must be taken with model parameterization and interpretation. It is recommended that future users perform appropriate sensitivity analyses before proceeding with the modelling task.

3.2 CASE STUDY: PANČEVAČKI RIT, SERBIA

3.2.1 Introduction

Pančevački Rit is situated on the left bank of the River Danube, north of Belgrade, in Serbia, and partly covers the north-west suburban areas of Belgrade. The area (Figure 3.17) is bounded by the River Danube to the west, the River Tamiš to the east, and in the north by the Karaš channel which connects the two rivers. The lowland region is partly urbanized with an area of about 34,000 ha and is protected by 90 km of levees. Land elevation lies between 69 and 76 m above sea level.

For the purpose of water management, the whole catchment area is divided into seven sub-catchments, each connected to a pumping station (Figure 3.17). The total installed capacity of the pumping stations is $34 \, m^3/s$. There is also an extensive drainage channel network in the region. The total length of the drainage network is about 870 km and its channel density (channel length per unit drained area) is approximately 25 m/ha.

Water movement in the region is influenced by water levels in the surrounding rivers, meteorological conditions and groundwater levels. Water levels in the River Danube are a function of recent hydrological conditions in addition to the downstream boundary condition. The latter is controlled by the Đerdap dam together with the Đerdap power plant built at the international border between Serbia and Romania. There are constant economic pressures to raise water levels in the Đerdap reservoir to produce more energy. Figure 3.18 compares water level duration curves in the study area for the natural, present and planned flow regimes for the River Danube. The dam clearly influences water levels, which are maintained higher than planned for significant periods of time. Figure 3.18 also shows the cumulative land elevation levels of the Pančevački Rit region revealing that about one-third of the area is below the average water level of the Danube.

Cumulative precipitation in an average year is 684 mm, but there are significant differences between wet and dry years. For example, 1999 was extremely wet with

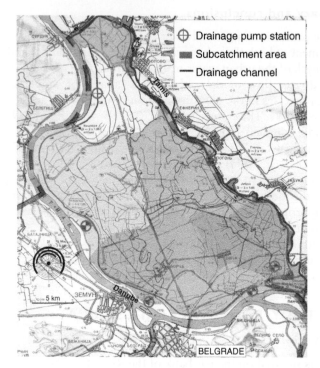

Figure 3.17 Pančevački Rit region: geographical location, sub-catchments and drainage channel network (see also colour plate 35)

Source: The authors

a cumulative precipitation of 1,049 mm, while the following year was extremely dry with a precipitation of only 367 mm. Such variability of hydrological conditions, along with very high water levels in the River Danube render water management in the area of Pančevački Rit very difficult. Groundwater levels are mainly influenced by the River Danube, as well as by a group of wells used to supply water to the town of Pančevo and to the food industry. The total capacity of the wells is about 500 l/s.

Meteorological conditions are another dominant influence on the River Danube regime. Table 3.6 shows monthly temperatures and average precipitation for the region.

Most of the land in the region is agricultural. It is estimated that about 25% of the region is urbanized with a strong tendency for rapid and mostly unplanned urban development. Such development further complicates control of the water regime, because the existing drainage system was designed using criteria suitable for arable lands, and it is not able to meet the needs of an urban environment.

The overall aim of the study presented here was to improve water management in the area of Pančevački Rit. The specific objectives were:

- to identify problems in the present water management,
- to define a broad strategy for general water management,
- to specify management criteria for each system (sewage system, channel network, pumping stations, abstraction wells),

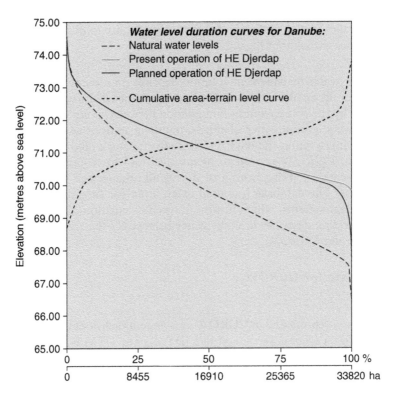

Figure 3.18 Water level duration curves for the River Danube in the Pančevački Rit study area: comparison of natural, present and planned regimes of the Đerdap dam. Also shown is the cumulative terrain level curve (land elevation curve) for the Pančevački Rit

Source: The authors

Table 3.6 Meteorological conditions in the study area

Average temperature (°C)

	JAN	FEB	MAR	APR	MAY	JUN	JUL	AUG	SEP	OCT	NOV	DEC	AV.
Average	0.7	2.7	7.0	12.3	17.3	20.5	22.2	22.0	17.8	12.5	7.0	2.6	12.0
Maximum	4.8	9.1	11.8	16.2	21.5	25	25.5	26.8	21.7	17.0	12.3	6.6	
Minimum	−5.5	−7.2	1.2	8.2	13.5	17.5	19.8	18.1	14.1	9.2	1.3	−1.9	
Standard deviation	2.42	3.42	2.66	1.79	1.77	1.52	1.37	1.76	1.58	1.51	2.30	2.25	

Average precipitation (mm)

	JAN	FEB	MAR	APR	MAY	JUN	JUL	AUG	SEP	OCT	NOV	DEC	SUM
P (mm)	45.2	40.2	44.4	55.9	69.4	94.9	69.2	50.6	54.6	45.6	55.4	58.6	684.2
(%)	6.60	5.88	6.49	8.17	10.14	13.87	10.11	7.40	7.98	6.66	8.10	8.56	100.0

- to evaluate the water balance, and
- to design a monitoring system to be used in the future to observe changes in the water balance throughout the area.

The quality of a water management solution depends on a reliable evaluation of the water balance for all significant water system components. The area of Pančevački Rit is characterized by the complexities of interaction between several different water systems, namely rivers, groundwater, channels, pumping stations and abstraction wells. Compiling and interpreting the data for all these systems would normally be a very challenging task. UGROW proved to be an ideal tool for coping with such a task because it is based around the concept of storing all urban water systems data within a single database. The database was used as an input for simulating the interaction between individual systems, which in turn allowed calculation of the water balance. The following section describes how the water balance for the area of Pančevački Rit was evaluated using UGROW.

3.2.2 Input data for UGROW

Terrain data

The land surface is represented in UGROW as a three-dimensional surface, mathematically described as a Digital Terrain Model (DTM). The DTM is generated from the (x, y, z) coordinates of a series of surface points called elevation points. Elevation points covering the area of Pančevački Rit were collected by digitizing contour lines from scanned maps at a scale of 1:5000. About 85,000 elevation points were collected in this way.

Geology data

In order to define the aquifer geometry, data from 155 boreholes were represented in the model. Figure 3.19 shows the DTM and the locations of the selected boreholes.

Once the elevations of the top and bottom of the aquifer were specified for each borehole, the aquifer geometry was defined by implementing the GEOSGEN algorithm for generating geology solids over the model area. Figure 3.20 shows the results of applying this algorithm for profiles P-1, P-2 and P-3 in the form of profiles of the land surface, aquifer top and aquifer bottom.

Drainage channel network

Three-dimensional data for the drainage channel network were entered into UGROW. The channels were digitized from scanned maps and checked by importing orthophoto images. Other data on the channel geometry, including cross-sectional areas, and upstream and downstream bed levels, were obtained from existing technical documentation.

Soil moisture balance

The period of simulation covers four years, from January 1999 to December 2002. Daily data for temperature, solar radiation, relative humidity and wind velocity were

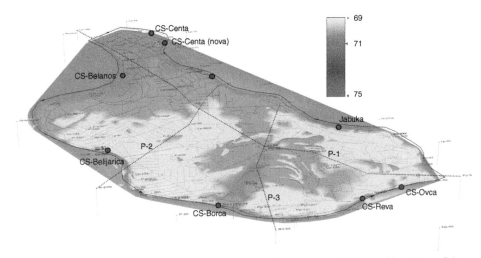

Figure 3.19 The Digital Terrain Model (DTM), locations of selected boreholes and locations of the cross-sections P-1, P-2 and P-3 shown in Figure 3.20 (see also colour plate 36)

Source: The authors

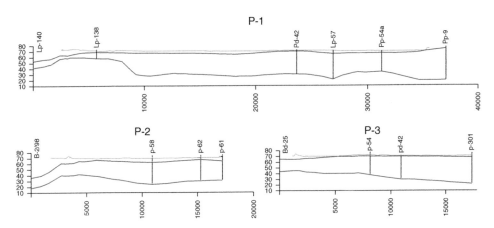

Figure 3.20 Aquifer geometry and Digital Terrain Model for the profiles P-1, P-2 and P-3. The locations of the profiles are given in Figure 3.19

Source: The authors

used to calculate potential evapotranspiration using the Penman-Monteith method. Subsequently, the UNSAT model was used to determine the soil moisture balance components, namely: runoff (Roff), leakage (Leak) and actual evapotranspiration (ET_a). Results for 2001 are shown in Figures 3.21 and 3.22.

Boundary conditions

Daily values of water level in the River Danube and River Tamiš were used to define the model boundary conditions. Interaction between these rivers and the aquifer

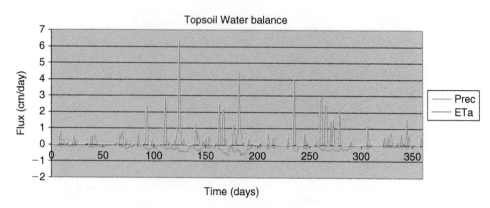

Figure 3.21 Precipitation and estimates of potential evapotranspiration as a result of implementing the *UNSAT* model for 2001

Source: The authors

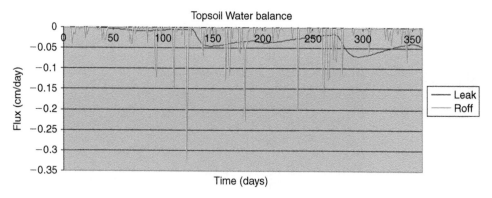

Figure 3.22 Estimates of leakage and runoff as a result of implementing the *UNSAT* model for 2001

Source: The authors

occurs via a sediment with a significantly lower hydraulic conductivity than the aquifer. In order to determine the characteristics of this layer, a well-known one-dimensional analytical model was introduced. This takes the form:

$$\frac{\partial^2 s}{\partial x^2} = \frac{S_y}{T} \frac{\partial s}{\partial t}$$

(3.2.1)

where, s is drawdown in the aquifer, x is distance from the river, S_y is specific yield, T is transmissivity, t is time.

and the boundary conditions are:

$$s(t = 0, x > 0) = 0$$
$$s(t > 0, x \to \infty) = s_0 \qquad (3.2.2)$$
$$\text{and, } s(t > 0, x = 0) = s_0$$

The analytical solution is:

$$s(x, t) = s_0(1 - \frac{2}{\sqrt{\pi}} \int_0^u e^{-u^2} du) = s_0 \cdot erfc(u) \qquad (3.2.3)$$

where u is a dimensionless variable $u = \sqrt{\dfrac{x^2 S}{4Tt}}$ and $erfc$ is the complementary error function.

For the daily fluctuation of water level in the river, the principle of superposition gives:

$$s(x, t) = \sum_i \Delta s_i \cdot erfc\left(\sqrt{\frac{x^2 S}{4T(t - t_i)}}\right) \qquad (3.2.4)$$

where Δs_i is the change in water level in the river between time t_i and the previous time t_{i-1}.

The model was calibrated using data from piezometers located close to the river bank, where it can be assumed that groundwater levels are dominated by water levels in the river. Figure 3.23 shows the results for piezometer CB-41, located near the Danube levees. The red line shows the measured water levels in the River Danube, the dotted line shows the blue line water level measurements in the piezometer for the same period, and the solid green line provides the simulation results. The specific recharge from the river was also calculated and is shown as a function of the water level in the Danube.

3.2.3 Simulation results

The modelling domain covered the whole study area shown in Figure 3.17, and was divided into finite elements by implementing the *MESHGEN+UFIND* algorithms. As explained above (Section 2.7), the *UFIND* algorithm determines the intersection in 3D between the drainage channel network and each finite element. Figure 3.24 shows the results of subdividing the modelling domain into finite elements.

Figure 3.24 also presents the results of the *DELINEATE* algorithm. As explained in Section 2.5, this algorithm subdivides the modelling area into sub-catchments for surface runoff simulation. As shown, each outlet (in this case the outlets are the drainage

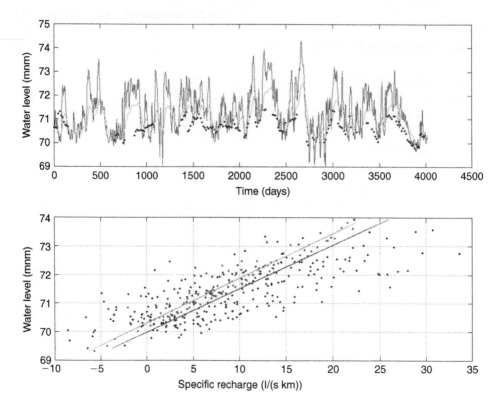

Figure 3.23 Results of the analytical 1D model (see also colour plate 37)

Source: The authors

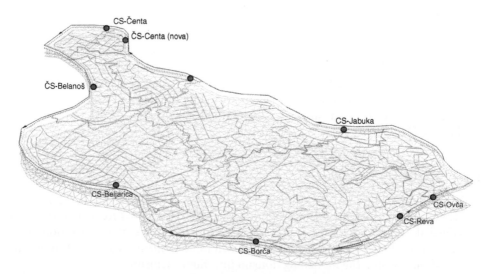

Figure 3.24 Results of mesh generation (green finite elements) and surface runoff delineation algorithms (catchment areas associated with each drainage outlet CS are shown with a dashed red line). Drainage channels are shown as blue lines (see also colour plate 38)

Source: The authors

pumping stations) has its own catchment area as defined by the network topology and the terrain elevation data for each element in the mesh.

The model was calibrated using:

- groundwater levels in selected piezometers, and
- measured discharge at outlets from the drainage network.

The average hydraulic conductivity in the region is 30 m/day. Figure 3.25 shows the transmissivity map.

The simulation period began in January 1999 and lasted four years until December 2002. Figure 3.26 shows the simulation results for 15 May 1999. Contour lines indicate groundwater levels and blue patches show poorly drained areas where the groundwater level is within 0.5 m of the ground surface. Because 1999 was an extremely wet year, a significant part of the area remained flooded despite the very dense drainage network. A major part of the poorly drained area is located on the left bank of the River Danube and is heavily urbanized.

Figure 3.26 also shows a comparison of field and model data. Simulated and measured discharge from the drainage pumping station at Reva is shown on the right-hand diagram. The simulated discharge is the sum of discharges calculated by the *RUNOFF* model and those contributed by groundwater as calculated by the *GROW* model. The *GROW* simulation also produces groundwater levels. The left-hand chart in Figure 3.26 compares the results of the *GROW* simulation with measurements in a selected piezometer.

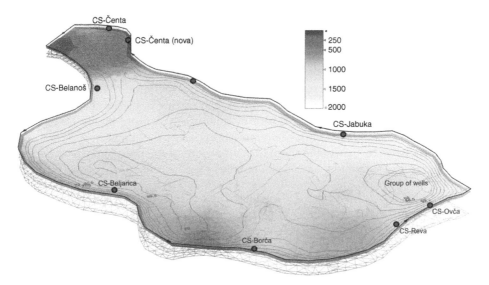

Figure 3.25 Aquifer transmissivity (m²/day) (see also colour plate 39)

Source: The authors

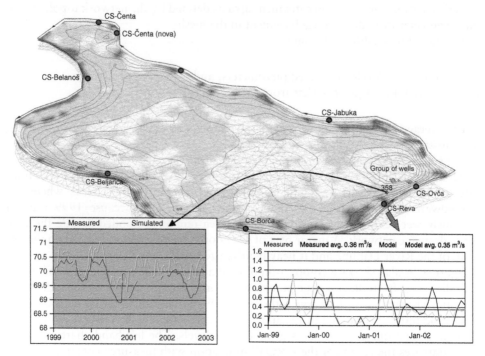

Figure 3.26 Charts showing simulation results for a selected piezometer and a drainage pumping station. Groundwater contours provide simulation results for 15 May 1999 (see also colour plate 40)

Source: The authors

3.2.4 Conclusion

Evaluating the water balance for a relatively large area with multiple land uses and a large number of water systems with complex geometries is a very challenging task. *UGROW* proved to be a powerful tool in this regard, storing highly diverse data within a single system, then graphically presenting and processing these data to produce a water balance. Bearing in mind the complexity of the task, agreement between the measured and simulated water balance results was excellent.

3.3 CASE STUDY: CITY OF BIJELJINA IN BOSNIA

3.3.1 Introduction

The city of Bijeljina is located in the province of Semberia, Bosnia (Figure 3.27). For at least ten years prior to the 1992 war, groundwater management in Bijeljina lacked clear objectives and a long-term strategy. The weakness of the groundwater management plan was revealed during the war, when groundwater resources were put under severe pressure due to rapid increases in the city population. The problems reached a peak

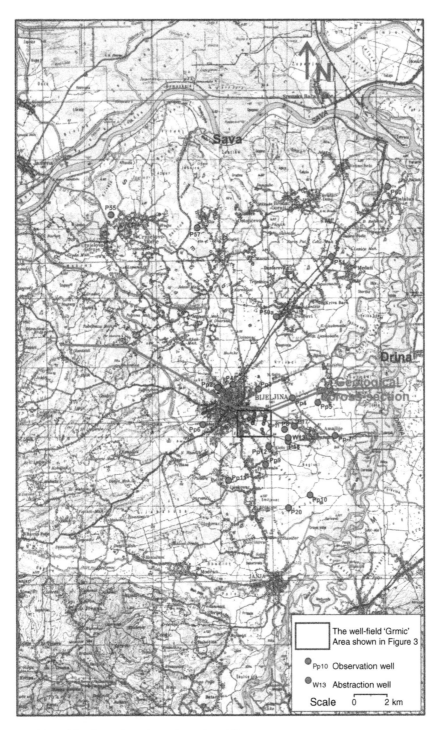

Figure 3.27 Map of Semberia (see also colour plate 41)

Source: After Pokrajac, 1999

when serious faecal coliform bacterial pollution was detected in the well water. In response, a project was initiated to define short-term and long-term groundwater management strategies. The objectives of the study were:

- to identify the source of bacterial pollution in the wells,
- to identify other potential sources of pollution, and
- to establish priorities for groundwater protection.

The probable source of the bacterial pollution was established, but quantitative analysis was required to justify the expense required to resolve the problem. The study was carried out in unfavourable circumstances, with limited resources and time. The problem was severe and a quick solution was needed. The extent and nature of the problem is described below, along with details of how a *UGROW* groundwater simulation model acted as a valuable decision-support tool.

The city of Bijeljina is situated between two rivers – the River Drina to the east and the River Sava to the north (Figure 3.27). The whole region has abundant groundwater stored in an alluvial aquifer (units 2 and 3 on Figure 3.28) some $400\,km^2$ in area. The aquifer is highly permeable and extremely vulnerable to water quality impacts. Pollutants leaching into the water table are rapidly transported by groundwater flow. During development of the city's sanitary system and the selection of solid waste disposal sites, little consideration was given to the shallow aquifer system. As a result, there are numerous sources of potential groundwater pollution. Water quality problems began to emerge during the 1990s.

The public water supply of Bijeljina depends on groundwater. The well-field is located close to the city (Figures 3.27 and 3.29) and properly constructed large diameter wells yield 7 to 9 Ml/day with only 1 m to 2 m of drawdown. The whole of the city is serviced by public supply, but the sewage system is only partially constructed. In parts of the city lacking public sewers, waste is directed to poorly constructed septic

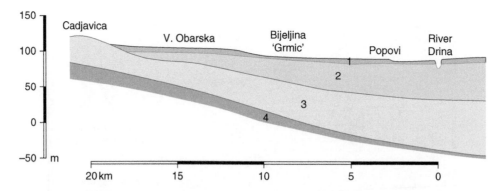

Figure 3.28 Representative west-east geological cross-section: (1) swamp clays; (2) sand and gravel; (3) sand, gravel with interbedded clays; (4) marl, marly clays (see also colour plate 42)

Source: After Pokrajac, 1999

tanks which, in turn, discharge to infiltration wells. This is the case along three streets close to the well-field (Figure 3.29):

- 'Hajduk Stanka' street which is within 150 m of well 10,
- 'Galac' street with a series of illegally built houses parallel to a line of wells about 300 m up the hydraulic gradient, and
- 'S. Jovanovica' street which is west of 'Galac' and slightly more remote from the pumping wells.

During a site visit, numerous septic tank infiltration wells were documented (twenty-four in 'Hajduk Stanka' street, thirty-one in 'Galac' and ten in 'S. Jovanovica' street), as well as a basin full of septic waste in 'Galac' street (marked in Figure 3.29).

Before 1992, when the war started in Bosnia, all demand for potable water was satisfied. Average daily consumption of water was approximately 12.1 Ml/day, while the total capacity of all wells was 24.1 Ml/day. During the war, an influx of migrants increased the population of 50,000 by a further 30,000.

The water authorities decided that the well capacity should be increased to 28.5 Ml/day. Besides the increased number of inhabitants, higher water losses due to the poor maintenance of the pipeline system during the war probably contributed to increased demand. After a period of intensive exploitation, serious entero-bacterial pollution was detected in the western wells in the summer of 1993. As a first measure, five wells with a total capacity of about 7 Ml/day were excluded from the system, and a new

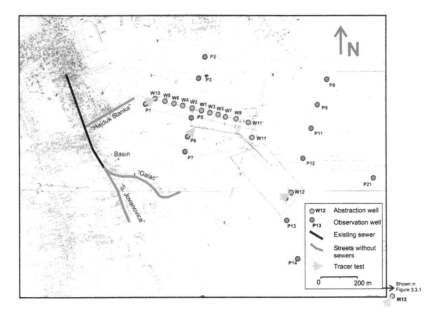

Figure 3.29 Layout map of the well-fields (see also colour plate 43)

Source: After Pokrajac, 1999

well with an equivalent capacity was constructed. This was just a temporary solution and a study was carried out to prioritize the water quality problems and recommend solutions.

3.3.2 Geology and hydrogeology

A representative geological cross-section (W-E) is shown in Figure 3.28. Geological units within the uppermost 200 m include:

- quaternary alluvial sediments, consisting of swamp clays (1) overlying a sand and gravel complex (2),
- paludin sand and sandy gravel with layers of clay (3),
- pontian marly clays and marl (4).

The main aquifer consists of alluvial sands and gravels deposited by the Drina and Sava Rivers (2) and the Paludin complex (3). These units are in very good hydraulic continuity and form a single aquifer system. The slope of the aquifer runs generally from the south-west to the north-east. Its thickness varies from 20 to 50 m in the western and southern parts of Semberia to between 90 and 120 m, even 170 m locally, in the northern and eastern parts close to the rivers. The alluvial gravel deposits are naturally protected from contamination by an overlying layer of clay, which has variable thickness, thinning and even disappearing in some areas. Consequently, groundwater resources are unprotected and prone to contamination in some localities.

3.3.3 Groundwater regime

Groundwater levels in Semberia are strongly influenced by the River Drina, which is in good hydraulic continuity with the aquifer. The River Drina recharges the aquifer along most of its length, the exception being the northern, lower part, close to its connection with the River Sava. Here the aquifer releases water to both the Drina and the Sava. Average precipitation in the annual, long-term water balance, is 781 mm, surface runoff is 219 mm, potential evapotranspiration is 522 mm and groundwater recharge from precipitation is about 120 mm. The principal direction of groundwater flow is from south to north, parallel to the River Drina. Seasonal fluctuations of groundwater are relatively moderate, with a maximum of 1.5 m close to the rivers, and 1 m in the zone of the well-field in Bijeljina.

3.3.4 Field measurements

Groundwater levels were monitored during the design, construction and operation of the well-field. However, the total number of monitoring wells and frequency of observation varied continually. During 1985 and 1986, groundwater levels were monitored weekly at thirty-six locations shown in Figures 3.27 and 3.29. Abstraction rates for the wells were estimated on the basis of head curves for the pumps which were calibrated only once with actual flow rates. After the construction of the large diameter wells 11, 12 and 13, standard pumping tests were performed, and their results were used to estimate the aquifer transmissivity.

To provide information on groundwater flow velocities, data from two previous tracer experiments (Avdagic, 1992) were analysed. In the first set of experiments, sodium chloride and a dye (sodium fluorescein) were used as tracers at two test sites shown in Figure 3.29. The tracers were released in up-gradient wells and monitored in down-gradient wells. Concentrations of sodium chloride were estimated by measuring the electrical resistance of the water samples. All the sampling wells were shallow, capturing water only from the uppermost 3 m of the aquifer. The results of the tracer studies are shown in Figure 3.30. The upper diagrams show measured dye concentrations, and the lower diagrams show measured electrical resistance values. Groundwater levels in the wells were not recorded. The average linear flow velocity ('pore velocity') was estimated at 7 m/day at location P6, but was not ascertained for location P1, where the time for maximum tracer concentration was not reliably determined.

In the second tracer study, pumping wells were used as sampling points, and tracers (dye) were introduced into neighbouring observation wells. This approach provided control over the groundwater flow direction (towards the well), and groundwater flux (a function of the well abstraction rate) during the experiment. The depth of the aquifer engaged in the test was the same as in the normal operation of the wells, so the information obtained during the tests can be considered representative of the behaviour of the aquifer as a whole. The tests were carried out for wells 12 and 13 (Figure 3.29). Prior to each test, a routine pumping test (step test) was performed to estimate the aquifer transmissivity. At the end of the step test, the final flow rate was maintained constant for several hours while introducing dye into the observation well. Water samples from the abstraction well were monitored until the first sign of dye was observed. This time was recorded as the travel time between the observation well and the abstraction well. The distance between the observation well and the abstraction well, the well discharge, the aquifer transmissivity (obtained from the pumping test) and the travel time are shown for each test in Table 3.7.

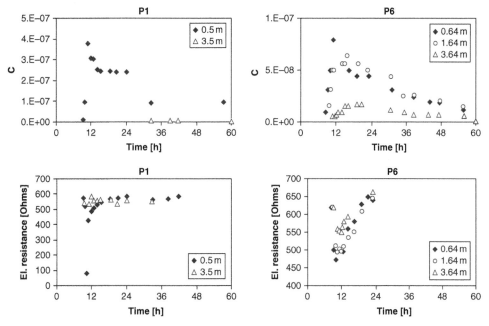

Figure 3.30 Break-through curves

Source: After Pokrajac, 1999

Table 3.7 Tracer test data for pumping wells W12 and W13

Quantity	Notation	Unit	W12	W13
Distance	r_o	m	10.8	39.5
Well discharge	Q	l/s	110	115
Transmissivity	T	m²/s	0.34	0.29
Travel time	t	h	1.3	12.5

For the purpose of the analysis, the following assumptions were made:

- flow towards the well is steady state and radial,
- the aquifer is confined with a constant thickness M, hydraulic conductivity K, effective porosity n_{eff} and aquifer transmissivity $T = KM$.

Under these assumptions, the travel time, τ, from the point the tracer is released to the point at distance r_o from the well, can be determined by integrating the pore velocity along a radius r, such that:

$$\tau = n_{eff} \, M\pi \frac{r_o^2 - r_w^2}{Q} \tag{3.3.1}$$

or

$$\frac{K}{n_{eff}} = \frac{T\pi}{\tau} \frac{r_o^2 - r_w^2}{Q} \tag{3.3.2}$$

where r_w is the radius of the well.

This equation allows the ratio of hydraulic conductivity to effective porosity, K/n_{eff}, to be calculated. It provides values for the two wells that are closely similar – 0.27 m/s (W12) and 0.28 m/s (W13). When realistic estimates of effective porosity are introduced, the hydraulic conductivity is estimated to be about 0.05 to 0.07 m/s. Such a high value of conductivity is not supported by the pumping test analyses and cannot be reconciled with model calibrations as discussed below. Instead, the tracer tests were interpreted to represent transport at maximum velocity, via the most permeable sub-layer. According to geological logs of the wells, this is a layer of coarse gravel. Since the study concerned transport of pathogenic bacteria in groundwater, the maximum velocity, albeit through a thin layer, is a highly relevant parameter.

In effect, the travel time observed reflects the hydraulic conductivity of the most permeable sub-layer, while the hydraulic conductivity obtained from pumping tests and calibrated model calibration represents the average over the full aquifer thickness:

$$K = \frac{\sum K_i m_i}{\sum m_i}, \quad \sum m_i = M, \tag{3.3.3}$$

where K_i is the hydraulic conductivity of the i-th sub-layer and m_i is its thickness. Vertically averaged hydraulic conductivity is smaller than the maximum hydraulic conductivity by definition, and can be much smaller if the vertical variability of hydraulic conductivity is significant. It is the vertical variability of hydraulic conductivity that causes the vertical variation in unit flux and this, in turn, is responsible for the longitudinal dispersion at the reservoir scale.

3.3.5 Urban aquifer model

GROW was used to develop an urban aquifer model of Semberia. The purpose of the modelling was to provide an assessment of the pollution sources and to establish priorities for groundwater protection. Due to the high permeability of the aquifer in Semberia, transport of pollutants is dominated by advection. Advective transport was modelled by particle tracking. Since data on groundwater quality were unavailable, modelling of transverse dispersion was not attempted.

The conceptual model of the aquifer in Semberia contains two layers – a high permeability aquifer consisting of geological units (2) and (3) with an impermeable base, and a low permeability confining layer comprising geological unit (1) (Figure 3.28). The numerical model was developed using a grid of 4,266 elements and 12,517 nodes as shown in Figure 3.31. The advantages of the finite-element technique were exploited in this study. Using curvilinear elements, the naturally irregular aquifer boundaries (the Drina and Sava rivers, and the aquifer's western limit) were represented realistically with very few approximations. Also, by refining the mesh in the vicinity of the well-field, the potentially problematical development and coupling of two separate models – a regional model and a local model – was avoided. The resulting model is large scale and regional but with a refined local zone of interest (zoomed circle in Figure 3.31), which ensures no significant loss of accuracy or waste of computational time.

The model was calibrated using data from 1985, when the number of operational observation wells was greatest. The average values for November 1985 were used for calibration because groundwater levels were steady during the whole month. Zones with similar geological features were represented by single values of hydraulic conductivity (Figure 3.31). Abstraction rates for pumping wells operating during November 1985 are given in Table 3.8. Water levels in the Sava and Drina rivers were obtained from gauging stations, and were interpolated for the sites between them. The values of hydraulic conductivity were found by trial and error, the objective being to minimize the difference between observed and calculated groundwater levels. In the vicinity of the wells, the values of hydraulic conductivity specified in the model were benchmarked against data obtained from pumping tests. The distribution of hydraulic conductivity values finally adopted is shown in Figure 3.31, and the measured and calculated values of groundwater levels for November 1985 are shown in Figure 3.32.

Once calibrated, the groundwater model was used to simulate the groundwater flow regime:

- prior to the detection of pollution in 1993, when all the wells were in operation, and
- immediately after the closure of the five western wells.

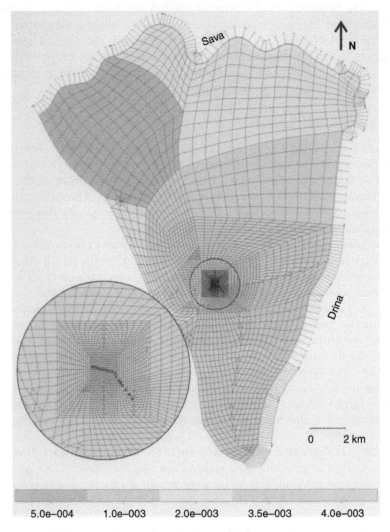

5.0e–004　　1.0e–003　　2.0e–003　　3.5e–003　　4.0e–003

Figure 3.31 Numerical grid and spatial distribution of hydraulic conductivity (Units m/s) (see also colour plate 44)

Source: After Pokrajac, 1999

Since data for 1993 were missing (except for estimated abstraction rates), data for 1994 were used. The abstraction rates for the period before and after the closure are given in Table 3.8. Regional groundwater levels and pathlines are shown in Figure 3.33; and Figure 3.34 compares capture zones and travel times for the two modelled periods. During the period prior to the closure of the wells, groundwater was moving from the area of Hajduk Stanka, Galac and S. Jovanovica streets towards the western wells (Figure 3.34, upper picture). As shown in the figure, pathlines from Hajduk Stanka miss well no. 10; however, since pathlines from several septic tanks located a short distance

Table 3.8 **Abstraction rates in ml/day for pumping wells in the model**

Well number	November 1985	Immediately before closure of polluted wells (data from 1994)	Immediately after closure of polluted wells (data from 1994)
10	1.4	1.4	–
8	1.4	1.4	–
6	1.4	1.4	–
4	1.4	1.4	–
2	1.4	1.4	–
1	1.4	1.4	1.4
3	1.4	1.4	1.4
5	1.4	1.4	1.4
7	4.3	4.3	4.3
9	4.3	4.3	4.3
11	4.3	4.3	4.3
11'	–	4.3	4.3
12	–	–	6.9
Totals	24.1	28.4	28.5

south east of Hajduk Stanka intercept the well, this street was included as a potential source of pollution. Closure of the five western wells displaces the capture zone to the east, and the groundwater pathways from numerous septic tanks along Galac and some along Hajduk Stanka miss the active wells (Figure 3.34, lower diagram).

It must be recognized that the results of the simulation incorporate a number of significant uncertainties. Firstly, the groundwater model was calibrated using a limited amount of data. The model needs to be recalibrated when more data are available to reduce uncertainty in the model parameters. Secondly, model results show only the average flow pattern for the 'average' abstraction rates provided by the water authorities. The actual situation in the summer of 1993 may have been worse with well capture zones temporarily larger due to periods of more intensive abstraction. Finally, in addition to the advective transport simulations depicted in Figure 3.34, some polluted water will probably move perpendicular to the pathlines due to lateral dispersion.

In order to use the regional (vertically-averaged) numerical model to calculate minimum travel times, effective porosity was artificially lowered by a factor of 5 to account for the results of the tracer tests. The corresponding travel times are shown in Figure 3.34. Before closure of the western pumping wells, some infiltration wells along the southern side of Hajduk Stanka street were shown to lie within ten days or less travel time of the pumping wells. The travel time for septic tank water contained in the Galac street basin is about twenty days.

The closure of the wells affects the flow pattern beneath Hajduk Stanka street, but has minimal influence on the flow beneath the two other streets. Therefore, the closure of the wells serves only as a short-term remedial strategy, and a more reliable plan is needed for the long term. Urgent construction of mains sewerage was proposed for the three streets and a network of observation wells was added to the existing one to monitor water quality and improve the reliability of the groundwater model. A recalibrated model will be used to assess the feasibility of reintroducing the western wells following

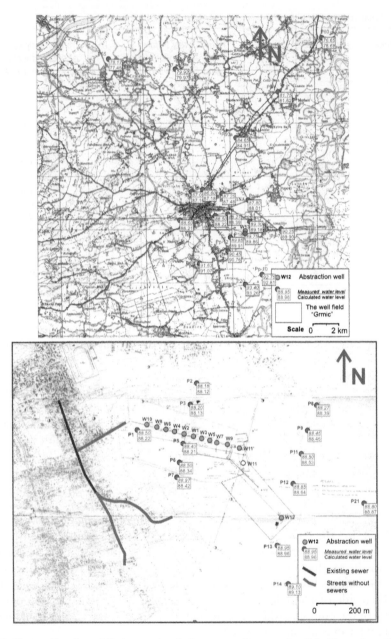

Figure 3.32 Measured and modelled water levels (in metres above sea level) in November 1985 (see also colour plate 45)

Source: After Pokrajac, 1999

the construction of mains sewerage. The operation of these wells would be based on water quality monitoring, with the possibility that wells closest to the streets would be used as interception wells, that is, for the active protection and control of groundwater quality. In addition to improved water level and quality monitoring, measurement of

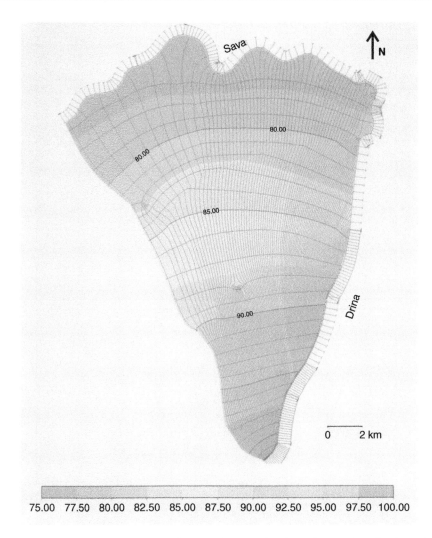

Figure 3.33 Modelled groundwater levels (metres above sea level) and flow paths in Semberia for 1994 (see also colour plate 46)

Source: After Pokrajac, 1999

pumping rates and volumes and the application of these data to model recalibration would assist greatly in addressing issues related to model non-uniqueness.

3.3.6 Concluding discussion

The results of a numerical groundwater flow model and *in situ* tracer experiments were used to confirm the source of bacterial pollution of groundwater abstracted to supply the city of Bijeljina with potable water. Model simulations demonstrated that while recharge from septic tanks had negligible influence on the groundwater regime, its impact on groundwater quality was significant and unacceptable. Estimated travel

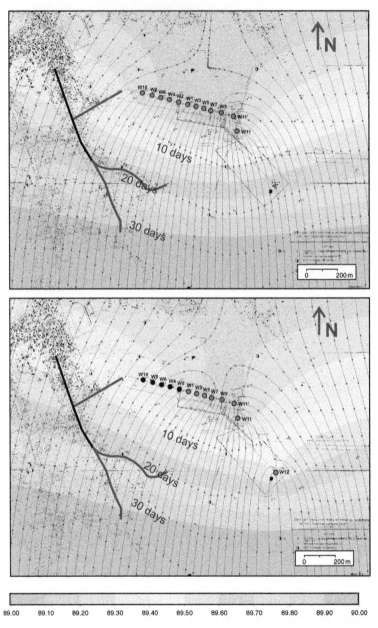

Figure 3.34 Capture zones and travel times before and after the closure of five western wells. Shading shows water levels in metres above sea level (see also colour plate 47)

Source: After Pokrajac, 1999

times from streets with septic tanks to the pumping wells ranged from three weeks to just a few days. Over this short period of time, bacteria entering the groundwater were able to survive and emerge in the well-water. The contamination arose due to a lack of mains sewerage in close vicinity to the well-field, a problem that resulted from a water management system that separated responsibilities for groundwater, water supply and wastewater, even though all three were operated by the same company. The study showed that domestic septic tanks in the vicinity of the well-field needed to be replaced by a public sewerage system.

The problem was caused by a failure to recognize the strong interrelationships that exist between municipal water systems. Water supply, urban drainage, solid waste and urban groundwater must be managed integrally. Water supply in the absence of a proper sewerage system potentially leads to the pollution of groundwater, and although it is impossible to prevent pollution completely, the release of contaminants underlying aquifers must be constrained to below critical levels. To achieve this goal, the decision-making processes for all municipal water systems need to be coordinated.

Reliable delineation of aquifer protection zones is a first step to protecting ground-water resources. Numerical models simulating the groundwater flow regime can be a powerful tool for decision-making. However, to avoid serious errors, they must be supported by appropriate field experiments. This is particularly true for simulations of contaminant transport, as opposed to simulations that only consider groundwater flux. Models that are unable to represent rapid flow through discrete high permeability layers, may give reliable predictions of groundwater levels and flux, but may produce erroneous predictions of pore velocities and travel times.

Even with the most advanced monitoring system, data are collected at isolated locations, and the model often interpolates between them. Thus, all model results are inherently uncertain. The uncertainties can be reduced step-by-step if the model is developed in stages, using the results of each stage to improve the monitoring system, collect more data, recalibrate the model and challenge the conclusions made thus far. With such a strategy, models can clarify our understanding of the processes that take place in complex municipal environments and provide very useful support for decision-making.

Chapter 4

Conclusions

Dubravka Pokrajac[1] and Ken W.F. Howard[2]

[1]School of Engineering, University of Aberdeen, Aberdeen, United Kingdom
[2]Department of Physical and Environmental Sciences, University of Toronto, Toronto, Canada

4.1 THE URBAN SUSTAINABILITY CHALLENGE

Ensuring healthy and sustainable living conditions in intensively populated environments has emerged as a major global challenge. The provision of safe and sustainable water supplies for drinking and sanitation is central to this undertaking. Historically, the vital role groundwater plays in the urban water cycle has suffered serious neglect. To some extent this reflects an 'out of sight, out of mind' mentality, which has promoted ignorance of water movement in the subsurface. However, neglect has also arisen because groundwater and surface water systems are spatially distinct and, in terms of water flow velocities, operate on totally different time scales. Reasons aside, the unfortunate consequence is that tools for urban water management rarely, if ever, incorporate an adequate understanding of urban aquifers and the role of groundwater, either during the analysis stage or, just as importantly, during the subsequent decision-making process. These attitudes must change and time is of the essence. Holistic management of the entire urban water cycle is starting to be recognized worldwide as an important priority. In turn, practical, soundly developed urban water system modelling tools are essential if the goal of urban sustainability is ever to be achieved.

Over the past twenty-five years, our understanding of urban groundwater issues and our ability to model them have both advanced significantly. However, developments in these areas have tended to progress independently and it is only in recent years that the paths have merged with serious consideration now given to designing models that explicitly incorporate features that are common to urban areas such as multiple point, line and distributed sources of contamination and the augmentation of aquifer recharge via sewers and leaking water supply networks. In this regard, AISUWRS (Assessing and Improving Sustainability of Urban Water Resources and Systems) currently represents one of the most comprehensive and sophisticated urban water modelling systems developed to date, and benefits from its scientific rigour, its attention to detail and thorough field testing. Unfortunately, the model's heavy data demands has inhibited its routine adoption by many of the world's cities. This problem is compounded by its 'coupling approach', which places strong reliance on its ability to link efficiently with independently developed groundwater flow models (e.g. Modflow or Feflow®) to complete the multi-component modelling package and achieve its ultimate objective.

4.2 *UGROW* AS A TOOL FOR URBAN WATER SYSTEM MANAGEMENT

As a fully integrated model dedicated to urban water systems, *UGROW* provides a complete and seamless modelling package that strongly complements AISUWRS in the suite of tools now available to urban water resource decision managers. *UGROW* may also offer certain advantages. Similar to AISUWRS, *UGROW* has been developed to support decision-making within the framework of integrated urban groundwater management, and its features have been designed to deal directly with some of the more important hydrogeological characteristics of the urban subsurface. It lacks some of the sophistication of AISUWRS, and in its current embryonic form has several limitations. Nevertheless, the inclusion of a dedicated, aquifer simulation module (*GROW*), which meshes smoothly and flawlessly with other urban water system model components, makes *UGROW* particularly attractive.

The original purpose of the *UGROW* software system was to raise awareness of the nature of interactions between urban groundwater and other urban water systems, and to improve the capability of simulation models to express these interactions. Thus, a main focus of *UGROW* has to been to create a tool that not only simulates urban water system interactions from a quantitative standpoint, but can also demonstrate and display these interactions in a highly visual way. To fulfil this task, a considerable amount of data describing various urban water systems needs to be stored and efficiently manipulated. To achieve this goal, *UGROW* has been developed with strong GIS functionality and fully integrates appropriate dynamic simulation models.

The main components of *UGROW* are its database, a suite of simulation models and the graphical user interface:

- The database contains all data on the geometry of the geological layers, the properties of the hydrogeological units and the hydraulic characteristics of the various elements of the urban water system. Its primary components are *TERRAIN*, *GEOLOGY* and *WATER*. *TERRAIN* is dedicated to the manipulation and presentation of the ground surface, *GEOLOGY* handles the geological layers, while *WATER* deals with the operating water systems such as streams, sewers and water supply networks. For groundwater simulation, *WATER* also defines the model domain boundaries and the hydrogeological units (the primary aquifer and, where present, an overlying aquitard); generates the finite-element mesh using the algorithm *MESHGEN*; and connects urban water networks to the groundwater simulation model using the algorithm *UFIND*.
- The suite of simulation models includes *RUNOFF*, which calculates and distributes surface runoff; *UNSAT* to represent seepage in the unsaturated zone and determine aquifer recharge; and *GROW* for groundwater flow. The groundwater flow module is the heart of the system and is fully integrated with other *UGROW* components. The model uses the finite-element approach and allows transient simulation of flow and contaminant behaviour in urban aquifers including the dynamic interactions that take place with other urban water systems.
- The user interface of *UGROW*, called *3DNet*, is an integrated hydro-informatics tool which has direct access to the *TERRAIN*, *GEOLOGY* and *WATER* components of the database. It is used primarily for
  - manipulating data,
  - step-by-step development of a site-specific model,
  - launching simulations, and
  - visualizing the results.

All information during the model development and subsequent simulations can be viewed as three-dimensional or two-dimensional graphics via the main *3DNet* window.

4.3 VALIDATION AND TESTING OF *UGROW*

Early versions of *UGROW* were tested using three urban groundwater case studies:

- Rastatt, Germany, where the main problem is infiltration of groundwater into sewers, increasing loads on the water treatment plant,
- Pančevački Rit, Serbia, where the urban water balance is poorly understood due to the presence of a large number of contributing water systems, and
- Bijeljina in Bosnia, where groundwater has been seriously polluted by subsurface emissions from septic tanks.

In each case *UGROW* performed successfully, the feedback from each study being used to refine model components. During the Rastatt study, the comparison of *UGROW* with the AISUWRS model suite demonstrated reasonably good agreement. Moreover, validation with the commercially distributed FEFLOW® simulation software was positive. Users not involved in the code development process were able to operate the *UGROW* system successfully, provided that appropriate support was available. All the studies showed that care must be taken with model parameterization and interpretation, and that appropriate sensitivity analyses should be performed before proceeding with the modelling task. At present, the main limitation of *UGROW* is its ability to simulate only one aquifer system. While this could be a problem where cities overlie multi-layer aquifer systems, much can be achieved by focusing on the uppermost aquifer since, as a general rule, protecting the uppermost aquifer automatically ensures that deeper aquifers are afforded comparable protection.

4.4 *UGROW* – THE FUTURE

UGROW is a powerful urban water management tool that can be used to raise awareness of the interactions between urban water system components, support decision-making and solve a wide range of urban water problems. The model has a sound scientific footing, is computationally efficient and is supported by excellent graphics. Field testing of early versions of the model has demonstrated its substantial potential. In the future, the model will be refined to extend its applicability to a broader range of hydrogeological conditions while retaining both its scientific rigour and its ease of use. The threats posed by climate change to the sustainability of groundwater in coastal cities are a particular challenge; but, with the continued support of users, the best of *UGROW* may yet be to come.

References

Alekperov, A.B., Agamirzayev, R.Ch. and Alekperov, R.A. 2006. Geoenvironmental problems in Azerbaijan. J.H. Tellam, M.O. Rivett and R.G. Israfilov (eds) *Urban Groundwater Management and Sustainability*. Dordrecht, Holland, Springer, pp. 39–58. (NATO Science Series, 74).

Anders, R. and Chrysikopoulos, C.V. 2005. Virus fate and transport during artificial recharge with recycled water. *Water Resources Research* 41(10).

Anon. 2002. *London's Warming: the impacts of climate change on London*. London, Greater London Authority.

Anon. 2004. *Nanoscience and nanotechnologies: opportunities and uncertainties*. Report. London, The Royal Society & The Royal Academy of Engineering.

Atkinson, T.C. 2003. Discussion of 'Estimating water pollution risks arising from road and railway accidents' by R.F. Lacey and J.A. Cole. *Quarterly J. Engineering Geology and Hydrogeology*, 36, pp. 367–68.

Atkinson, T.C. and Smith, D.I. 1974. Rapid groundwater flow in fissures in the chalk: an example from South Hampshire. *Quarterly J. Engineering Geology*, Vol. 7, pp. 197–205.

Attanayake, P.M. and Waterman, M.K. 2006. Identifying environmental impacts of underground construction. *Hydrogeology J.*, Vol. 14, pp. 1160–70.

Avdagic, I. 1992. *Podzemne vode jugoistocnog dijela Semberije* (Groundwater in south-east part of Semberia), Sarajevo, Institute of Hydraulic Engineering, Faculty of Civil Engineering (in Serbian).

Barrett, M.H., Hiscock, K.M., Pedley, S.J., Lerner, D.N., Tellam, J.H. and French, M.J. 1999. Marker species for identifying urban groundwater recharge sources: the Nottingham case study. *Water Research*, Vol. 33, pp. 3083–97.

Bear, J. and Bachmat, Y. 1991. *Introduction to Modeling of Transport Phenomena in Porous Media*. Dordrecht/Boston, Kluwer Academic Publishers, p. 553.

Bradford, T. 2004. *The Groundwater Diaries: trials, tributaries and tall stories from beneath the streets of London*. London, Flamingo, Harper Collins.

Brassington, F.C. 1991. Construction causes hidden chaos. *Geoscientist*, Vol. 14, pp. 8–11.

Burn, S., Desilva, D., Ambrose, M., Meddings, S., Diaper, C., Correll, R., Miller R. and Wolf, L. 2006. A decision support system for urban groundwater resource sustainability. *Water Practice & Technology*, Vol. 1.

Burston, M.W., Nazari, M.M., Bishop, P.K. and Lerner, D.N. 1993. Pollution of groundwater in the Coventry region (UK) by chlorinated hydrocarbon solvents. *J. Hydrology*, Vol. 149, pp. 137–61.

Butler, D. and Davies, J.W. 2000. *Urban Drainage*. London, E & FN Spon.

Carlyle, H.F., Tellam, J.H. and Parker, K.E. 2004. The use of laboratory-determined ion exchange parameters in the prediction of field-scale major cation migration over a 40-year period. *Journal of Contaminant Hydrology*, Vol. 68, pp. 55–81.

Carsel, R.F. and Parrish, R.S. 1988. Developing joint probability distributions of soil water retention characteristics. *Water Resources Research*, 24(5), pp. 755–69.

Cedergren, H.R. 1989. *Seepage, Drainage, and Flow Nets*, 3rd edn. New York, John Wiley.

Chilton, P.J. (ed.) 1999. *Groundwater in the Urban Environment: selected city profiles*. Rotterdam, Holland, Balkema.

Chilton, P.J. et al. (eds) 1997. *Groundwater in the Urban Environment: Vol. 1: Problems, Processes and Management*. Proc. of the XXVII IAH Congress on Groundwater in the Urban Environment, Nottingham, UK, 21–27 September 1997. Rotterdam, Holland, Balkema.

Chocat, B. 1997. Amenagement urbain et hydrologie. *La Houille Blanche*, Vol. 7, pp. 12–19.

Cook, S., Vanderzalm, J., Burn, S., Dillon, P. and Page, D. 2006. A karstic aquifer system. *Urban Water Resources Toolbox Integrating Groundwater into Urban Water management*. Mount Gambier, Australia, IWA Publications.

Cronin, A.A., Rueedi, J., Joyce, E. and Pedley, S. 2006. Monitoring and managing the extent of microbiological pollution in urban groundwater systems in developed and developing countries. J.H. Tellam, M.O. Rivett and R.G. Israfilov (eds) *Urban Groundwater Management and Sustainability*. Dordrecht, Holland, Springer, pp. 299–314. (NATO Science Series, 74.)

Datry, T., Malard, F. and Gibert, J. 2006. Effects of artificial stormwater infiltration on urban groundwater ecosystems. J.H. Tellam, M.O. Rivett and R.G. Israfilov (eds) *Urban Groundwater Management and Sustainability*. Dordrecht, Holland, Springer, pp. 331–45. (NATO Science Series, 74).

DeSilva, D., Burn, S., Tjandraatmadja G., Moglia, M., Davis, P., Wolf, L., Held, I., Vollersten, J., Williams, W. and Hafskjold, L. 2005. Sustainable management of leakage from wastewater pipelines. *Water Science and Technology*, 52(12), pp. 189–98.

Diaper, C. and Mitchell, G. 2006. Urban Volume and Quality (UVQ). L. Wolf, B. Morris and S. Burn (eds) *Urban Water Resources Toolbox: Integrating Groundwater into Urban Water Management*. London, IWA, pp. 16–33.

Diersch, H.-J-G. 2005. WASY Software FEFLOW, Finite Element Subsurface Flow and Transport Simulation Software, Reference Manual. Berlin, WASY, Institute for Water Resources Planning and Systems Research.

Dillon, P.J. and Pavelic, P. 1996. *Guidelines on the Quality of Stormwater and Treated Wastewater for Injection into Aquifers for Storage and Reuse*. Melbourne, Urban Water Research Association of Australia. (Research Report No. 109).

Eiswirth, M. 2002. Hydrogeological factors for sustainable urban water systems. K.W.F. Howard and R. Israfilov (eds) *Current Problems of Hydrogeology in Urban Areas, Urban Agglomerates and Industrial Centres*. Dordrecht/Boston: Kluwer Academic Publishers, pp. 159–84. (NATO Science Series IV, Earth and Environmental Sciences, 8).

Eiswirth, M., Ohlenbusch, R. and Schnell, K. 1999. Impact of chemical grout injection on urban groundwater. B. Ellis (ed.) *Impacts of Urban Growth on Surface and Groundwater Quality* pp. 187–94. (IAHS Pub. No. 259).

Eiswirth, M., Wolf, L. and Hötzl, H. 2004. Balancing the contaminant input into urban water resources. *Environmental Geology*, 46(2), pp. 246–56.

Ellis, P.A. and Rivett, M.O. 2006. Assessing the impact of VOC-contaminated groundwater on surface water at the city scale. *J. Contaminant Hydrology*, Vol. 91, pp. 107–27.

Ford, M. and Tellam, J.H. 1994. Source, type of extent of inorganic contamination within the Birmingham urban aquifer system, UK. *J. Hydrology*, Vol. 156, pp. 101–35.

Ford, M., Tellam, J.H. and Hughes, M. 1992. Pollution-related acidification in the urban aquifer, Birmingham, UK. *J. Hydrology*, Vol. 140, pp. 297–312.

Foster, S., Morris, B., Lawrence, A. and Chilton, J. 1999. Groundwater impacts and issues in developing cities – an introductory review. P.J. Chilton (ed.) *Groundwater in the Urban Environment*. Rotterdam, Holland, Balkema, pp. 3–16.

Fram, M.S. 2003. *Processes Affecting the Trihalomethane Concentrations Associated with the Third Injection, Storage and Recovery Test at Lancaster, Antelope Valley, California,*

March 1998 through April 1999. US Geological Survey Water Resources Investigations Report 03-4062.

Garcia-Fresca B. 2007. Urban-enhanced groundwater recharge: review and case. Study of Austin, Texas, USA. K.W.F. Howard (ed.) *Urban Groundwater – Meeting the Challenge*. London, Taylor & Francis, pp. 3–18. (IAH-SP Series, Vol. 8).

Gerber, R.E. 1999. Hydrogeologic behaviour of the Northern Till aquitard near Toronto, Ontario. Ph.D. thesis, Toronto, Ontario, University of Toronto.

Gerber, R.E. and Howard, K.W.F. 1996. Evidence for recent groundwater flow through Lake Wisconsinan till near Toronto, Ontario. *Canadian Geotechnical Journal*, Vol. 33, pp. 538–55.

Gerber, R.E. and Howard, K.W.F. 2000. Recharge through a regional till aquitard: three dimensional flow model water balance approach. *Groundwater*, Vol. 38, pp. 410–22.

Gerber, R.E. and Howard, K.W.F. 2002. Hydrogeology of the Oak Ridges Moraine aquifer system: implications for protection and management from the Duffins Creek watershed. *Canadian Journal of Earth Sciences* (CJES), Vol. 39, pp. 1333–48.

Glass, R.J., Steenhuis, T.S. and Parlange, J.-Y. 1988. Wetting front instability as a rapid and far-reaching hydrologic process in the vadose zone. *J. Contaminant Hydrology*, Vol. 3, pp. 207–26.

Global Water Partnership. 2000. *Integrated Water Resources Management*. TAC Background Papers, No. 4, www.gwpforum.org/gwp/library/Tacno4.pdf

Global Water Partnership. 2002. *ToolBox, Integrated Water Resources Management*. http://gwpforum.netmasters05.netmasters.nl/en/index.html

Grimmond, C.S.B. and Oke, T.R. 1999. Evapotranspiration rates in urban areas. B. Ellis (ed.) *Impacts of Urban Growth on Surface and Groundwater Quality*, IAHS Pub. No. 259, pp. 235–44.

Harris, J.M. 2007. Precipitation and urban runoff water quality in non-industrial areas of Birmingham, UK. Unpublished MPhil Thesis. Birmingham, UK, University of Birmingham, Earth Sciences.

Harrison, R.M. and de Mora, S.J. 1996. *Introductory Chemistry for the Environmental Sciences*. 2nd edn. Cambridge, UK, Cambridge University Press.

Heathcote, J.A., Lewis, R.T. and Sutton, J.S. 2003. Groundwater modelling for the Cardiff Bay Barrage, UK – prediction, implementation of engineering works and validation of modelling. *Quarterly J. Engineering Geology and Hydrogeology*, Vol. 36, pp. 159–172.

Held, I., Wolf, L., Eiswirth, M. and Hötzl, H. 2006. Impacts of sewer leakage on urban groundwater. J.H. Tellam, M.O. Rivett and R.G. Israfilov (eds) *Urban Groundwater Management and Sustainability*. Dordrecht, Holland, Springer, pp. 189–204. (NATO Science Series, 74).

Hiscock, K.M. and Grischek, T. 2002. Attenuation of groundwater pollution by bank filtration. *Journal of Hydrology*, Vol. 266, pp. 139–44.

Hosseinipour, E.Z. 2002. Managing groundwater supplies to meet municipal demands: the role of simulation-optimization-demand models and data issues. J.H. Tellam, M.O. Rivett and R.G. Israfilov (eds) *Urban Groundwater Management and Sustainability*. Dordrecht, Holland, Springer, pp. 137–56. (NATO Science Series, 74).

Howard, K.W.F. 1988. Beneficial aspects of sea-water intrusion. *Ground Water*, Vol. 17, pp. 250–57.

Howard, K.W.F. (ed.) 2007. *Urban Groundwater – Meeting the Challenge*. London, Taylor & Francis. (IAH-SP Series, Vol. 8).

Howard, K.W.F. and Beck, P.J. 1993. Hydrogeochemical implications of groundwater contamination by road de-icing chemicals. *Journal of Contaminant Hydrology* 12(3), pp. 245–68.

Howard K.W.F. and Gelo, K. 2002. Intensive groundwater use in urban areas: the case of megacities. R. Llamas and E. Custodio (eds) *Intensive use of Groundwater: Challenges and Opportunities*. Rotterdam, Holland, Balkema, pp. 35–58.

Howard, K.W.F. and Haynes, J. 1993. Urban Geology 3: Groundwater contamination due to road de-icing chemicals – salt balance implications. *Geoscience Canada*, 20(1), pp. 1–8.

Howard, K.W.F. and Israfilov, R.G. 2002. *Current Problems of Hydrogeology in Urban Areas, Urban Agglomerates and Industrial Centres.* Dordrecht/Boston: Kluwer Academic Publishers. (NATO Science Series IV, Earth and Environmental Sciences Vol. 8).

Howard K.W.F. and Maier H. 2007. Road de-icing salt as a potential constraint on urban growth in the Greater Toronto Area, Canada. *Journal of Contaminant Hydrology*, Vol. 91, pp. 146–70.

Jaiprasart, P. 2005. Analysis of interaction of urban water systems with groundwater aquifer: a case study of the Rastatt City in Germany. Masters Thesis, Imperial College, London.

Johanson, R.C., Imhoff, J.C., Kittle Jr., J.L. and Donigian, A.S. 1984. *Hydrological Simulation Program – FORTRAN (HSPF): Users Manual for Release 8.0.* Athens, GA, Environmental Research Laboratory, US EPA.

Jones, H.K., MacDonald, D.M.J. and Gale, I.N. 1998. *The Potential for Aquifer Storage and Recovery in England and Wales.* British Geological Survey Report WD/98/26.

Jones, I., Lerner, D.N. and Thornton, S.F. 2002. A modelling feasibility study of hydraulic manipulation: a groundwater restoration concept for reluctant contaminant plumes. S.F. Thornton and S.E. Oswald (eds) *Groundwater Quality: natural and enhanced restoration of groundwater pollution.* pp. 525–31. (IAHS Publication No. 275).

Joyce, E., Rueedi, J., Cronin, A., Pedley, S., Tellam, J.H. and Greswell, R.B. 2007. *Fate and Transport of Phage and Viruses in UK Permo-Triassic Sandstone Aquifers.* Environment Agency of England and Wales Science Report SC030217/SR.

Khazai, E. and Riggi, M.G. 1999. Impact of urbanization on the Khash aquifer, an arid region of southeast Iran. B. Ellis (ed.) *Impacts of Urban Growth on Surface and Groundwater Quality*, pp. 211–17. (IAHS Publication No. 259).

Klinger, J. and Wolf, L. 2004. *Using the UVQ Model for the Sustainability Assessment of the Urban Water System.* EJSW workshop on process data and integrated urban water modelling. Proceedings availabile online at: http://www.citynet.unife.it/

Klinger, J., Wolf, L. and Hötzl, H. 2005. New modelling tools for sewage leakage assessment and the validation at a real world test site. *Proceedings of EWRA2005 – Sharing a Common Vision for our Water Resources.* 6th International Conference, 7–10 Sept. Menton, France.

Klinger, J., Wolf, L., Schrage, C., Schaefer, M. and Hötzl, H. A porous aquifer: Rastatt. L. Wolf, B. Morris and S. Burn (eds) *Urban Water Resources Toolbox: Integrating Groundwater into Urban Water Management.* London, IWA, pp 100–43.

Knipe, C., Lloyd, J.W., Lerner, D.N. and Greswell, R.B. 1993. *Rising Groundwater Levels in Birmingham and their Engineering Significance.* Special Publication 92. London, Construction Industry Research and Information Association (CIRIA).

Krothe, J.N. 2002. Effects of urbanization on hydrogeological systems: the physical effects of utility trenches. MS thesis, The University of Texas, Austin.

Krothe, J.N., Garcia-Fresca, B. and Sharp Jr., J.M. 2002. Effects of urbanization on groundwater systems. Abstracts for the International Association of Hydrogeologists XXXII Congress, Mar del Plata, Argentina, p. 45.

Kung, K.-J.S. 1990. Preferential flow in sandy vadose zone: 2. mechanisms and implications. *Geoderma*, Vol. 46, pp. 59–71.

Lacey, R.F. and Cole, J.A. 2003. Estimating water pollution risks arising from road and railway accidents. *Quarterly Journal of Engineering Geology and Hydrogeology*, Vol. 36, pp. 185–92.

Lerner, D.N. 1986. Leaking pipes recharge groundwater. Ground Water, 24(5), 654–662.

Lerner, D.N. 1997. Too much or too little: recharge in urban areas. P.J. Chilton et al. (eds) *Groundwater in the Urban Environment: Problems, Processes and Management.* Rotterdam, Holland, A.A. Balkema, pp. 41-47.

Lerner, D.N. 2002. Identifying and quantifying urban recharge: a review. *Hydrogeology J.*, Vol. 10, pp. 143–52.

Lerner, D.N. (ed.) 2003. *Urban Groundwater Pollution.* London, Taylor and Francis. (International Association of Hydrogeologists International Contributions to Hydrogeology ICH24).

Lerner, D.N. and Tellam, J.H. 1992. The protection of urban groundwater from pollution. *J. Institution of Water and Environmental Management*, Vol. 6, pp. 28–37.

Lerner, D.N., Yang, Y., Barrett, M.H. and Tellam, J.H. 1999. Loadings of non-agricultural nitrogen in urban groundwater. B. Ellis (ed.) *Impacts of Urban Growth on Surface and Groundwater Quality*. pp. 117–24. (IAHS Publication No. 259).

Maidment, D.R. 1993. *Handbook of Hydrology*. New York, McGraw-Hill.

Martin, P., Turner, B., Dell, J., Payne, J., Elliot, C. and Reed, B. 2001. *Sustainable Drainage Systems – Best Practice Manual for England, Scotland, Wales and Northern Ireland*. London, CIRIA, p 113. (CIRIA Report C523).

McDonald, M.G. and Harbaugh, A.L. 1988. *A Modular Three-Dimensional Finite-Difference Ground-water Flow Model: U.S. Geological Survey Techniques of Water-Resources Investigations*, book 6, chap. A1.

McDonald, M.G. and Harbaugh, A.W. 2003. The history of MODFLOW. *Ground Water*, 41(2), pp. 280–83.

Misstear, B., White, M., Bishop, P. and Anderson, G. 1996. Reliability of sewers in environmentally vulnerable areas. *Construction Industry Research and Information Association (CIRIA) Project Report 44*. London, CIRIA.

Mitchell, V.G. and Diaper, C. 2005. UVQ: A tool for assessing the water and contaminant balance impacts of urban development scenarios. *Water Science and Technology*, 52(12), pp. 91–98.

Morris, B.L., Lawrence, A.R. and Foster, S.D. 1997. Sustainable groundwater management for fast growing cities: mission achievable or mission impossible? J. Chilton et al. (eds) *Proceedings of the 27th IAH Congress on groundwater in the urban environment: problems, processes and management*. Rotterdam, Balkema, pp. 55–66.

Morris, B., Rueedi, J. and Mansour, M. 2006. A sandstone aquifer: Doncaster, UK. L. Wolf, B. Morris and S. Burn (eds) *Urban Water Resources Toolbox: Integrating Groundwater into Urban Water Management*. London, IWA.

Pedley, S. and Howard, G. 1997. The public health implications of microbiological contamination of groundwater. *Quarterly Journal of Engineering Geology and Hydrogeology*, 30(2), pp. 179–88.

Petts, G., Heathcote, J. and Martin, D. 2002. *Urban Rivers: Our Inheritance and Future*. London, IWA.

Pitt, R., Clark, S. and Field, R. 1999. Groundwater contamination potential from stormwater infiltration practices. *Urban Water*, 1(3), pp. 217–36.

Pokrajac, D. 1999. Interrelation of wastewater and groundwater management in the city of Biljeljina in Bosnia, *Urban Water*, Vol. 1, pp. 243–55.

Powell, K.L., Barrett, M.H., Pedley, S., Tellam, J.H., Stagg, K.A., Greswell, R.B. and Rivett, M.O. 2000. Enteric virus detection in groundwater using a glasswool trap. O. Sililo (ed.) *Groundwater: Past Achievements and Future Challenges*. Rotterdam, Balkema, pp. 813–16.

Powell, K.L., Taylor, R.G., Cronin, A.A., Barrett, M.H., Pedley, S., Sellwood, J., Trowsdale, S.A. and Lerner, D.N. 2003. Microbial contamination of two urban sandstone aquifers in the UK. *Water Research*, Vol. 37, pp. 339–52.

Preene, M. and Brassington, R. 2003. Potential groundwater impacts from civil-engineering works. *Water and Environment Journal*, 17(1), pp. 59-64.

Pyne, R.D.G. 2005. *Aquifer Storage Recovery: A Guide to Groundwater Recharge Through Wells*, Gainesville, FL, ASR Press.

Rivett, M.O., Lerner, D.N., Lloyd, J.W. and Clark, L. 1990. Organic contamination of the Birmingham aquifer. *Journal of Hydrology*, Vol. 113, pp. 307–23.

Rivett, M.O., Shepherd, K.A., Keys, L. and Brennan, A.E. 2005. Chlorinated solvents in the Birmingham aquifer, UK: 1986–2001. *Quarterly Journal of Engineering Geology and Hydrogeology*, 38(4), pp. 337–50

Robins, N.S., Kinniburgh, D.G. and Bird, M.J. 1997. Generation of acid groundwater beneath City Road, London. A.B. Hawkins (ed.) *Ground Chemistry Implications for Construction*.

Bristol, UK, Proc. Int. Conf. on the Implications of Ground Chemistry and Microbiology for Construction, 1992, pp. 225–32.

Rosenbaum, M.S., McMillan, A.A., Powell, J.H., Cooper, A.H., Culshaw, M.G. and Northmore, K.J. 2003. Classification of artificial (man-made) ground. *Engineering Geology*, Vol. 69, pp. 399–409.

Rueedi, J., Cronin, A.A., Moon, B., Wolf, L. and Hötzl, H. 2005. Effect of different water supply strategies on water and contaminant fluxes in Doncaster, United Kingdom. *Water Science and Technology*, 52(9), pp. 115–23.

Rushton, K.R. and Redshaw, S.C. 1979. *Seepage and Groundwater Flow*. Chichester, UK, Wiley.

Rushton, K.R., Kawecki, M.W. and Brassington, F.C. 1988. Groundwater model of conditions in the Liverpool sandstone aquifer. *J. Institution of Water and Environmental Management*, Vol. 2, pp. 65–84.

Scheytt, T., Grams, S. and Fell, H. 1998. Occurrence and behaviour of drugs in groundwater. J.V. Brahana, Y. Eckstein, L.K. Ongley, R. Schneider and J.E. Moore (eds) *Gambling with Groundwater – physical, chemical, and biological aspects of aquifer-stream relations*. St Paul, Minnesota, US, American Institute of Hydrology, pp. 13–18.

Seymour, K.J., Ingram, J.A. and Gebbett, S.J. 2006. Structural controls on groundwater flow in the Permo-Triassic sandstones of NW England. R.D. Barker and J.H. Tellam (eds) *Fluid Flow and Solute Movement in Sandstones: the Offshore UK Permo-Triassic Red Bed Sequence*. London, Geological Society, pp. 169–86. (Special Publications, 263).

Sharp Jr., J.M., Hansen, C.N. and Krothe, J.N. 2001. Effects of urbanization on hydrogeological systems: the physical effects of utility trenches. K-P. Seiler and S. Wohnlich (eds) *New Approaches Characterizing Groundwater Flow*, supplement volume. Proceedings of the International Association of Hydrogeologists XXXI Congress, Munich, Germany.

Sharp Jr., J.M., Krothe, J., Mather, J.D., Garcia-Fresca, B. and Stewart, C.A. 2003. Effects of urbanization on groundwater systems. G. Heiken, R. Fakundiny and J. Sutter (eds) *Earth Science in the City*. Washington, DC, American Geophysical Union, pp. 257–78. (Special Publication Series, 56).

Shepherd, K.A., Ellis, P.A. and Rivett, M.O. 2006. Integrated understanding of urban land, groundwater, baseflow and surface-water quality – The City of Birmingham, UK. *The Science of the Total Environment*, Vol. 360, pp. 180–95.

Smith, J.W.N. 2005. *Groundwater – Surface Water Interactions in the Hyporheic Zone*. Bristol, UK, Environment Agency. (Environment Agency Science Report SC030155/1).

Sophocleous, M. 2000. From safe yield to sustainable development of water resources: the Kansas experience. *Journal of Hydrology*, vol. 235, pp. 27–43.

Sophocleous, M. 2005. Groundwater recharge and sustainability in the high plains aquifer in Kansas, USA. *Hydrogeology Journal*, Vol. 13, pp. 351–65.

Sophocleous, M. 2007. Science and practice of environmental flows and the role of hydrogeologists. *Ground Water*, 45(4), pp. 393–401.

Souvent, P., Vizintin, G. and Cencur-Cuck, B. 2006. A layered aquifer system: Ljublana, Slovenia. L. Wolf, B. Morris and S. Burn (eds) *Urban Water Resources Toolbox: Integrating Groundwater into Urban Water Management*. London, IWA, pp. 191–216.

Taylor, R.G., Cronin, A.A., Lerner, D.N., Tellam, J.H., Bottrell, S.H., Rueedi, J. and Barrett, M.H. 2006. Hydrochemical evidence of the depth of penetration of anthropogenic recharge in sandstone aquifers underlying two mature cities in the UK. *Applied Geochemistry*, Vol. 21, pp. 1570–92.

Taylor, R.G., Cronin, A.A., Trowsdale, S.A., Baines, O.P., Barrett, M.H. and Lerner, D.N. 2003. Vertical groundwater flow in Permo-Triassic sediments underlying two cities in the Trent River Basin (UK). *J. Hydrology*, Vol. 284, pp. 92–113.

Tellam, J. H. and Thomas, A. 2002. Well water quality and pollutant source distributions in an urban aquifer. K.W.F. Howard and R.G. Israfilov (eds) *Current Problems in Hydrogeology in*

Urban Areas, Urban Agglomerates and Industrial Centres. Dordrecht, Holland, Springer, pp. 139–58. (NATO Science Series IV. Earth and Environmental Sciences, 8).

Tellam, J.H., Rivett, M.O. and Israfilov, R. (eds) 2006. *Urban Groundwater Management and Sustainability.* Dordrecht, Holland, Springer. (NATO Science Series, 74).

Thomas, A. and Tellam, J.H. 2006a. Modelling of recharge and pollutant fluxes to urban groundwaters. *Science of the Total Environment,* Vol. 360, pp. 158–79.

Thomas, A. and Tellam, J.H. 2006b. Development of a GIS model for assessing groundwater pollution from small scale petrol spills. K.W.F. Howard (ed.) *Urban groundwater – meeting the challenge.* Selected papers from the 23rd International Geological Congress, 2004, Florence, Italy. London, Taylor & Francis, pp. 107–27.

United Nations. 2005. *World Urbanization Prospects.* Department of Economic and Social Affairs. New York, UN.

Van Genuchten, M.T. 1980. A closed-form equation for predicting the hydraulic conductivity of unsaturated soils. *Soil Sci. Soc. Am. J.* Vol. 44, pp. 892–98.

Van Hofwegen, P.J.M. and Jaspers, F.G.W. 1999. *Analytical Framework for Integrated Water Resources Management. Guidelines for Assessment of Institutional Frameworks.* Monograph 2, Delft, Holland, Int. Inst. Hydraulic and Environmental Eng.

Van de Ven, F.H.M. and Rijsberman, M. 1999. Impact of groundwater on urban development in The Netherlands. J.B. Ellis (ed.) *Impact of Urban Growth on Surface water and Ground Water Quality.* Wallingford, UK, IAHS, pp. 13–21. (IAHS publication 259).

Wedjen, B. and Ovstedal, J. 2006. Contamination and degradation of de-icing chemicals in the unsaturated and saturated zones at Oslo Airport, Gardermoen, Norway. J.H. Tellam, M.O. Rivett and R.G. Israfilov (eds) *Urban Groundwater Management and Sustainability.* Dordrecht, Holland, Springer, pp. 205–18. (NATO Science Series, 74).

Wolf, L. 2004. Integrating leaky sewers into numerical groundwater models. EJSW Workshop on Process Data and Integrated Urban Water Modelling. Proceedings availabile online at: http://www.citynet.unife.it/

Wolf, L. 2006. Assessing the influence of leaky sewer systems on groundwater resources beneath the City of Rastatt, Germany. Ph.D. thesis, Department of Applied Geology. Karlsruhe, University of Karlsruhe.

Wolf, L. and Hötzl, H. 2006. Upscaling of laboratory results on sewer leakage and the associated uncertainty. K.F. Howard (ed.) *Urban Groundwater – Meeting the Challenge.* London, Taylor and Francis, pp. 79–94. (IAH selected paper series).

Wolf, L., Eiswirth, M. and Hötzl, H. 2006a. Assessing sewer-groundwater interaction at the city scale based on individual sewer defects and marker species distributions. *Environmental Geology,* Vol. 49, pp. 849–57.

Wolf, L., Morris, B. and Burn, S. (eds) 2006b. *AISUWRS Urban Water Resources Toolbox – Integrating Groundwater into Urban Water Management.* London, IWA.

Wolf, L., Held, I., Eiswirth, M. and Hötzl, H. 2004. Impact of leaky sewers on groundwater quality. *Acta Hydrochimica Hydrobiologie,* Vol. 32, pp. 361–73.

Wolf, L., Held, I., Klinger, J. and Hötzl, H. 2006c. Integrating groundwater into urban water management. *Water Science and Technology,* 54(6–7), pp. 395–403.

Wolf, L., Morris, B., Dillon, P., Vanderzalm, J., Rueedi, J. and Burn, S. 2006d. AISUWRS Urban Water Resources Toolbox – A Brief Summary. L. Wolf, B. Morris and S. Burn (eds) *AISUWRS Urban Water Resources Toolbox – Integrating Groundwater into Urban Water Management.* London, IWA, pp. 282–97.

Wolf, L., Klinger, J., Held, I., Neukum, C., Schrage, C., Eiswirth, M. and Hötzl, H. 2005. Rastatt City Assessment Report. AISUWRS Deliverable D9, Department of Applied Geology Karlsruhe. www.urbanwater.de

Yang, Y., Lerner, D.N., Barrett, M.H. and Tellam, J.H. 1999. Quantification of groundwater recharge in the city of Nottingham, UK. *Environmental Geology,* Vol. 38, pp. 183–98.

Index

Plate 1 A recent major urban landslide in Baku, Azerbaijan (site **X**), was directly associated with high
groundwater levels, caused by a combination of heavy rain and leaking water mains

Source: The authors

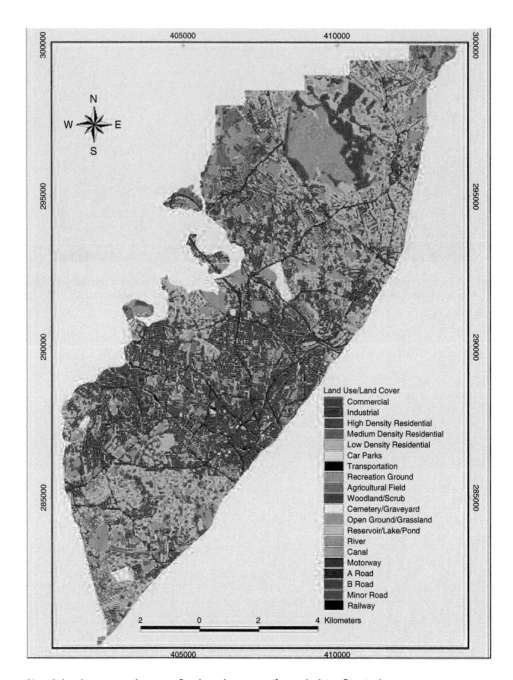

Plate 2 Land cover on the unconfined sandstone aquifer underlying Birmingham.

Source: After Thomas and Tellam, 2006b. This map is based in part on Ordnance Survey data: © Crown copyright Ordnance Survey.

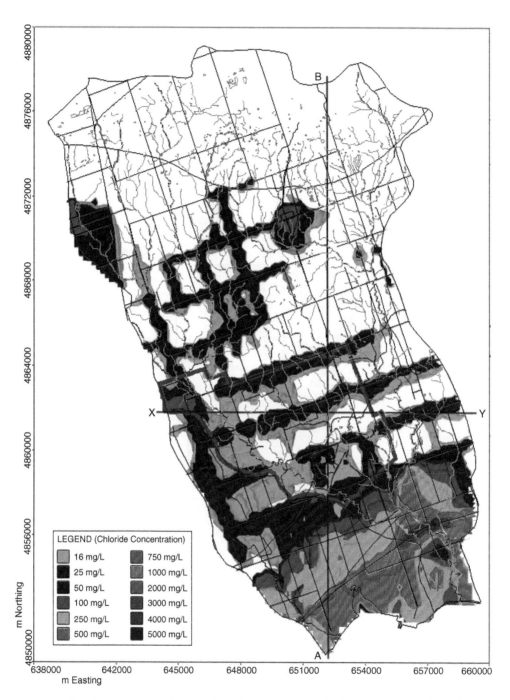

Plate 3 Predicted long-term, steady-state chloride concentrations in the uppermost aquifer in the absence of urban development in the Seaton Lands study area. Note that while chemical steady state is not achieved for several hundred years, most of the change occurs within a time-frame of about 100 years.

Source: The authors

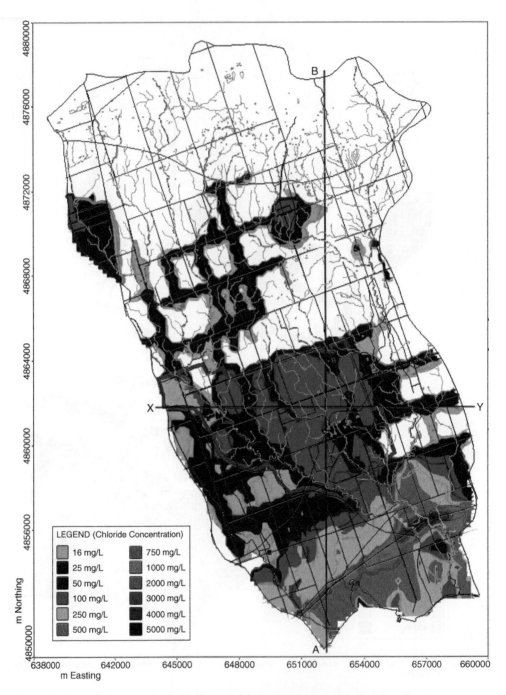

Plate 4 Predicted long-term, steady-state chloride concentrations in the uppermost aquifer as a result of road salt application following development of the Seaton Lands study area. Note that while chemical steady state is not achieved for several hundred years, most of the change occurs within a time-frame of about 100 years.

Source: The authors

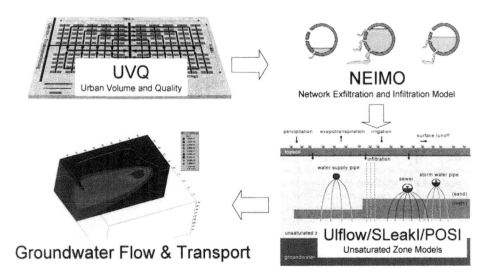

Plate 5 Major model compartments demonstrating the integrated approach of AISUWRS

Source: After Wolf et al, 2006b

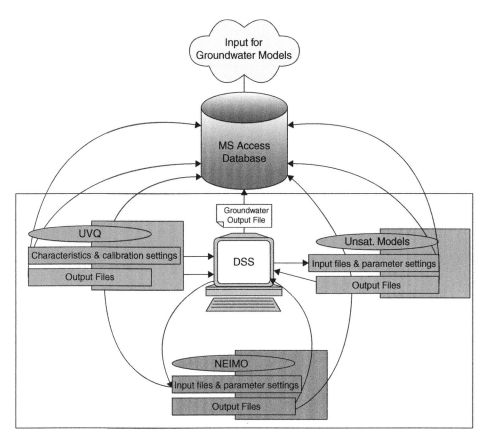

Plate 6 Link between the key AISUWRS model components, the decision support system (DSS) and the Microsoft Access database

Source: After Wolf et al, 2006b

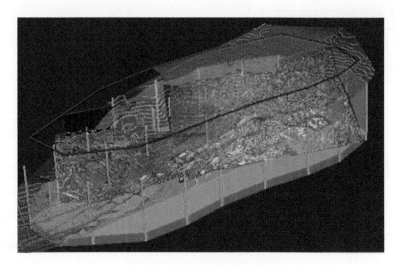

Plate 7 A 3D view of a terrain model and the hydrogeological layers

Source: The authors

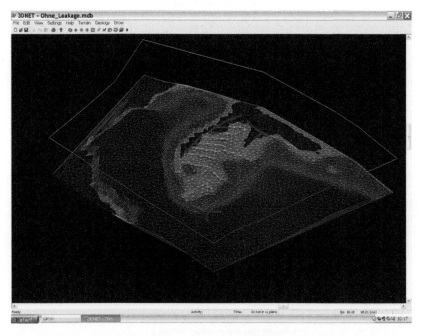

Plate 8 View of a water supply pipe and sewer in the city of Rastatt. The details of the case study are presented in Section 3.1

Source: The authors

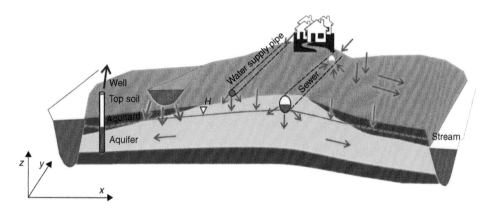

Plate 9 **A typical physical system that can be simulated using *UGROW* consists of a land surface with various land uses, an aquifer, upper and lower aquitards, an unsaturated zone, water supply mains, sewers, wells, streams and other urban water features.**

Source: The authors

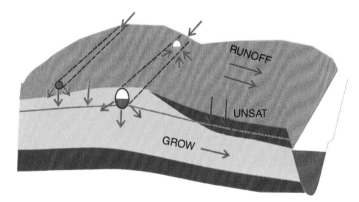

Plate 10 **The three simulation models related to physical processes in the urban water balance**

Source: The authors

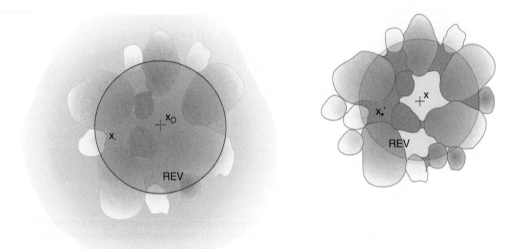

(Left) Plate 11 **Representative elementary volume in a saturated soil**

Source: The authors

(Right) Plate 12 **Representative elementary volume in unsaturated soil: blue = water, grey = air, brown = solids**

Source: The authors

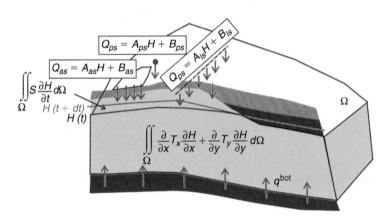

Plate 13 **Components of the aquifer water balance**

Source: The authors

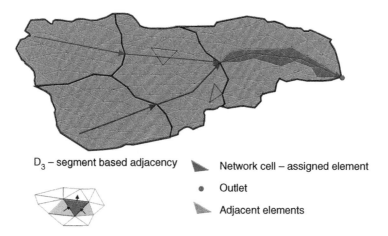

D₃ – segment based adjacency

Network cell – assigned element

Outlet

Adjacent elements

Plate 14 TIN-based delineation – D$_3$ propagation algorithm

Source: The authors

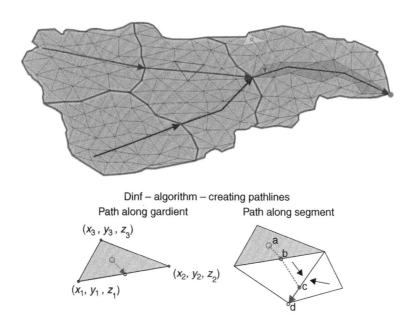

Dinf – algorithm – creating pathlines

Path along gardient

Path along segment

(x_3, y_3, z_3)

(x_2, y_2, z_2)

(x_1, y_1, z_1)

a
b
c
d

Plate 15 TIN-based delineation – D$_{inf}$ algorithm

Source: The authors

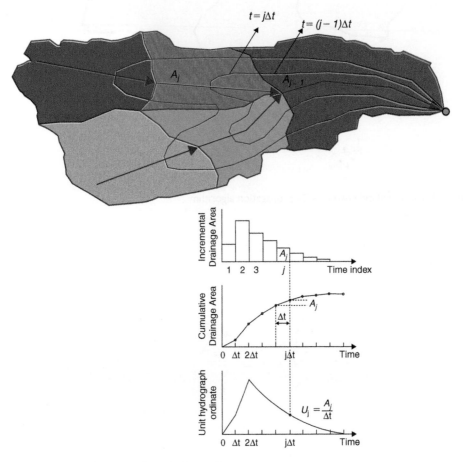

Plate 16 Time–area diagram and unit hydrograph

Source: The authors

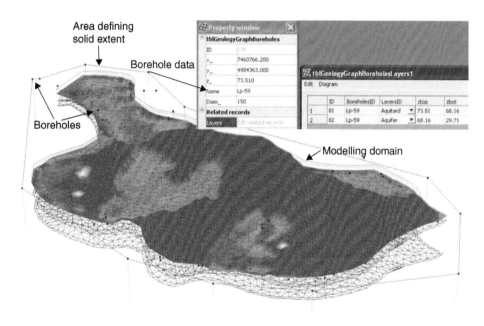

Plate 17 Geological layers in Pancevacki rit. The details of the case study are presented in Section 3.2

Source: The authors

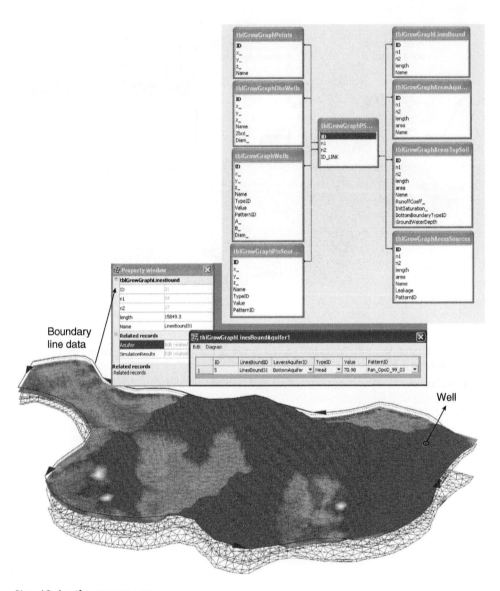

Plate 18 **Aquifer components**

Source: The authors

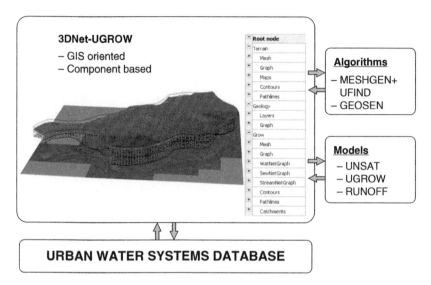

Plate 19 *3DNet-UGROW* and on-screen links

Source: The authors

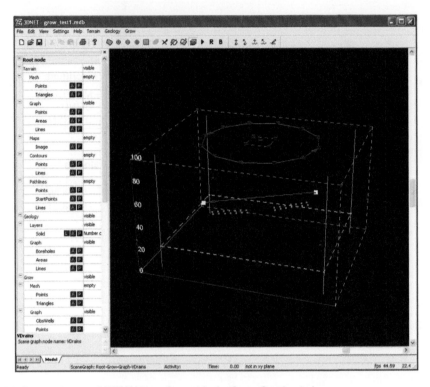

Plate 20 Starting layout of *UGROW* interface with the SceneGraph window

Source: The authors

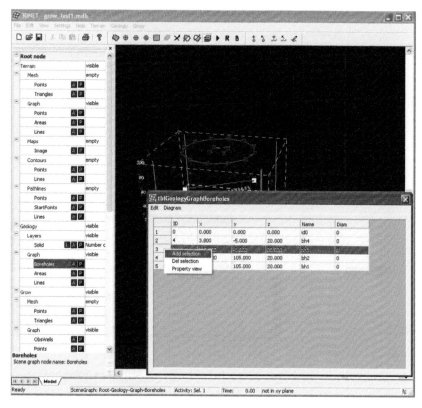

Plate 21 **Selecting objects from the grid attribute dialog box**

Source: The authors

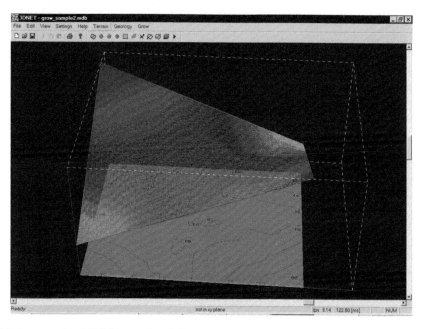

Plate 22 **An example of DTM created with Terrain ▶ Mesh triangulate** command

Source: The authors

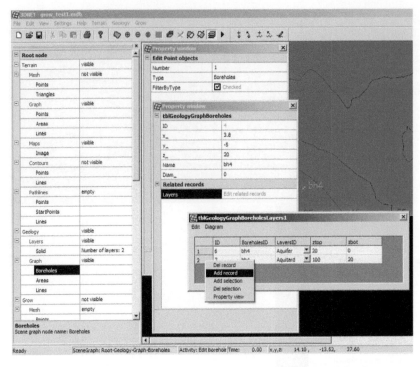

Plate 23 Assigning layers to boreholes via the Edit Point tool and Attributes dialog box

Source: The authors

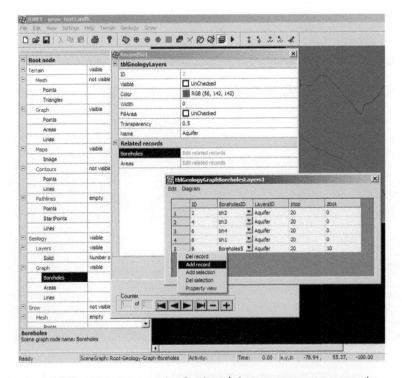

Plate 24 Assigning layers to boreholes via the **Geology ▶ Layer manager** command

Source: The authors

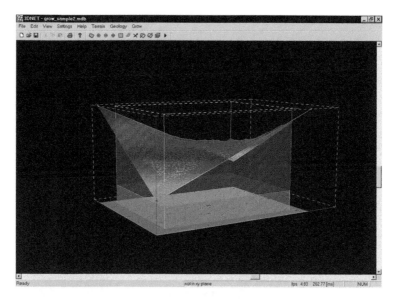

Plate 25 An example of two simple geological solids created from geology layers in four boreholes positioned at four corners of a rectangular area

Source: The authors

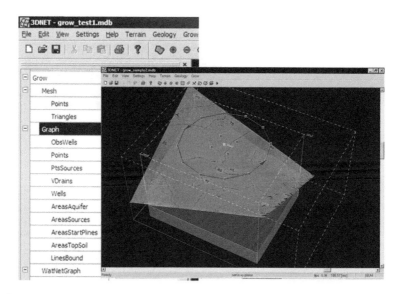

Plate 26 SceneGraph window showing all available objects under the **Grow-Graph** node and a model window showing an example of a boundary line for defining the modeling domain

Source: The authors

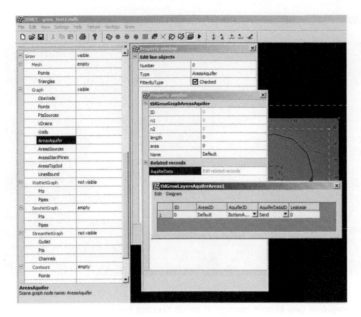

Plate 27 Assigning 'types' to the topsoil solid and aquifer solid

Source: The authors

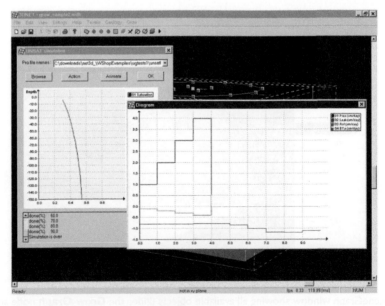

Plate 28 Results of a simulation using the *UNSAT* model

Source: The authors

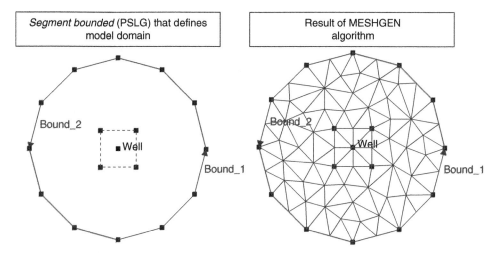

Plate 29 Triangulating the model domain

Source: The authors

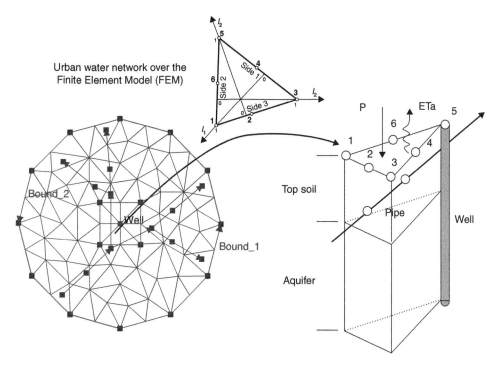

Plate 30 Defining vertical water balance input data for each mesh element

Source: The authors

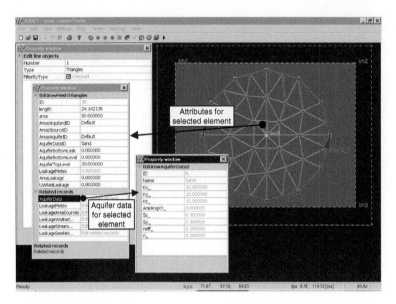

Plate 31 Editing a mesh element and viewing its attributes

Source: The authors

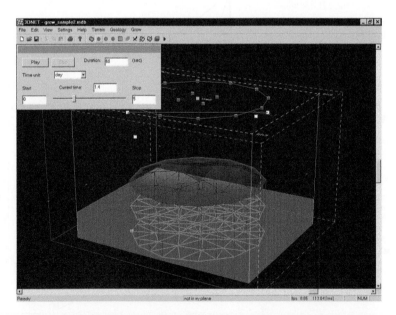

Plate 32 Simulation of groundwater flow affected by leakage from a water supply pipe and discharge from a well in the centre of the modeling domain

Source: The authors

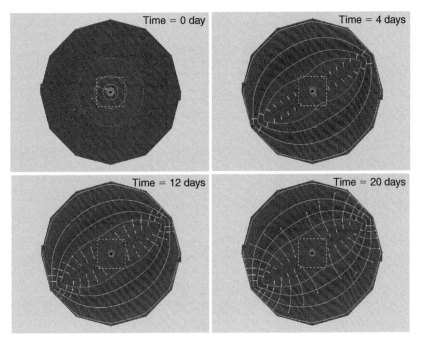

Plate 33 Results of implementing the pathline algorithm

Source: The authors

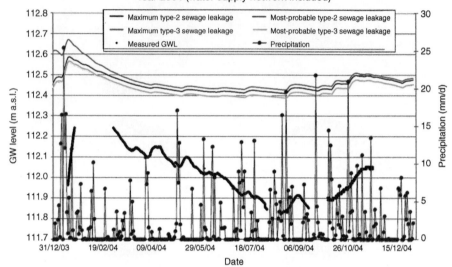

Plate 34 Comparison of modelled groundwater levels and measured data

Source: The authors

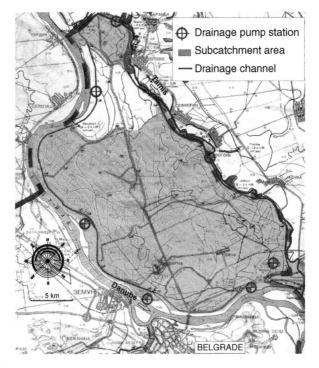

Plate 35 Pančevački Rit region: geographical location, sub-catchments and drainage channel network
Source: The authors

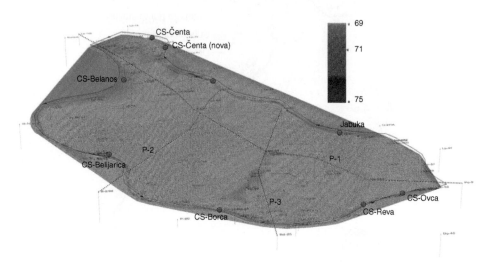

Plate 36 The Digital Terrain Model (DTM), locations of selected boreholes and locations of the cross-sections P-1, P-2 and P-3 shown in Figure 3.20

Source: The authors

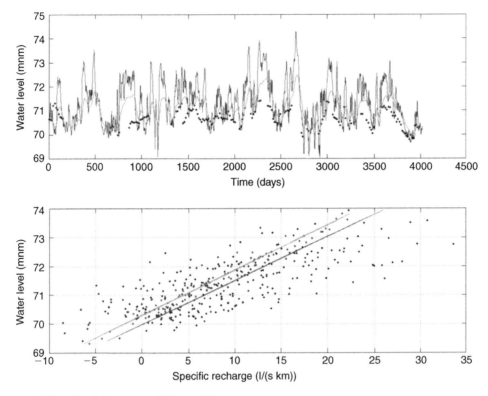

Plate 37 Results of the analytical 1D model

Source: The authors

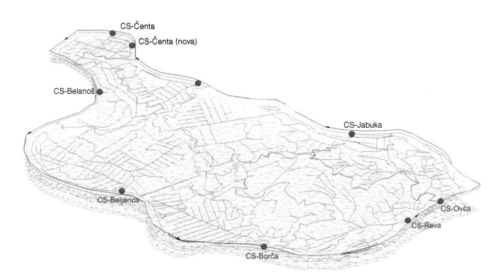

Plate 38 Results of mesh generation (green finite elements) and surface runoff delineation algorithms (catchment areas associated with each drainage outlet CS are shown with a dashed red line). Drainage channels are shown as blue lines

Source: The authors

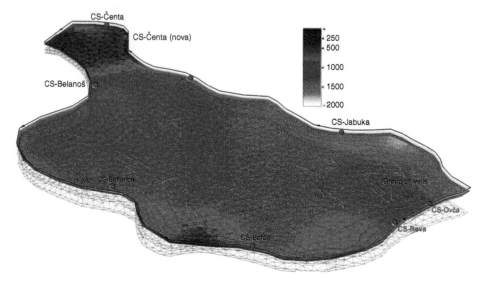

Plate 39 Aquifer transmissivity (m²/day)

Source: The authors

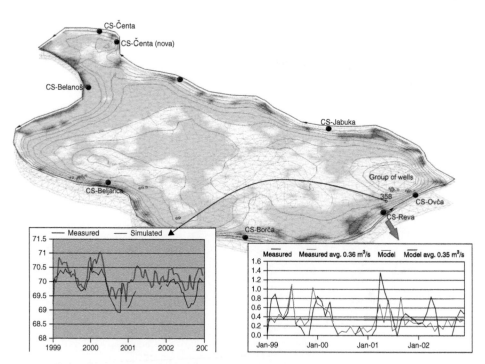

Plate 40 Charts showing simulation results for a selected piezometer and a drainage pumping station. Groundwater contours provide simulation results for 15 May, 1999

Source: The authors

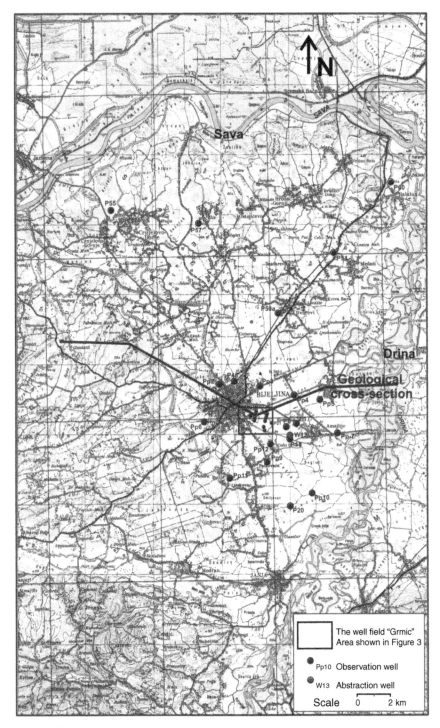

Plate 41 **Map of Semberia**

Source: After Pokrajac, 1999

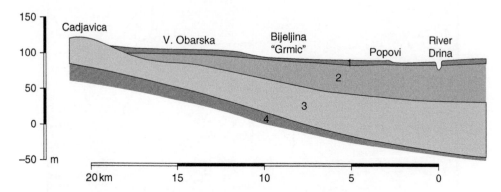

Plate 42 Representative west-east geological cross section: (1) swamp clays; (2) sand and gravel; (3) sand, gravel with interbedded clays; (4) marl, marly clays

Source: After Pokrajac, 1999

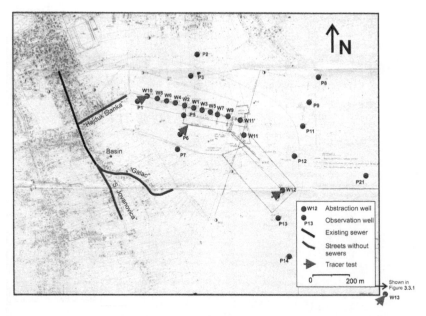

Plate 43 Layout map of the well fields

Source: After Pokrajac, 1999

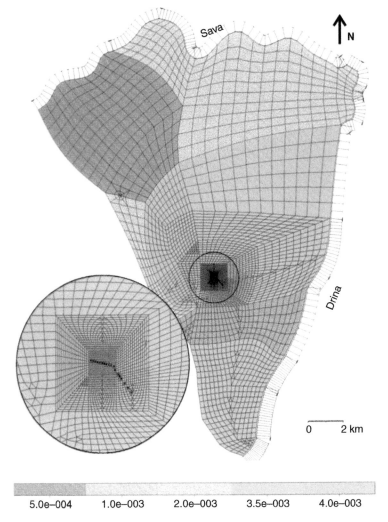

Plate 44 Numerical grid and spatial distribution of hydraulic conductivity (Units m/s)

Source: After Pokrajac, 1999

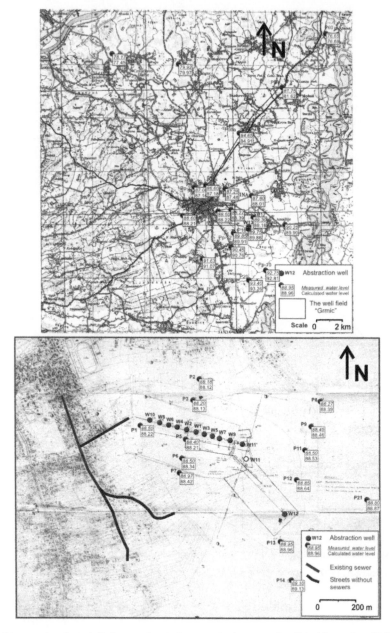

Plate 45 Measured and modelled water levels (in metres above sea level) in November 1985

Source: After Pokrajac, 1999

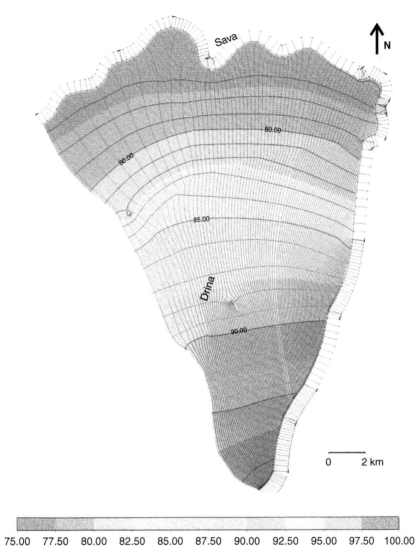

Plate 46 Modelled groundwater levels (metres above sea level) and flow paths in Semberia for 1994

Source: After Pokrajac, 1999

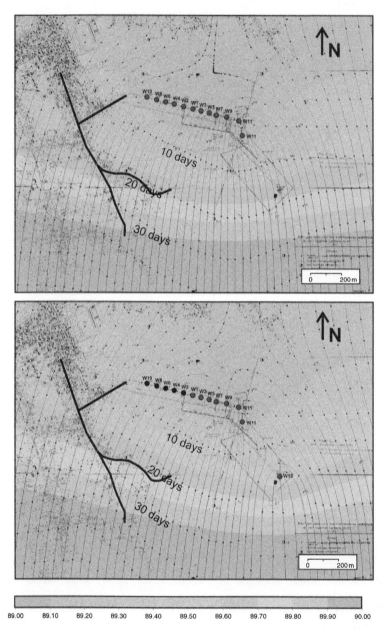

Plate 47 **Capture zones and travel times before and after the closure of five western wells. Shading shows water levels in metres above sea level**

Source: After Pokrajac, 1999

Printed and bound by CPI Group (UK) Ltd, Croydon, CR0 4YY

23/10/2024

01778257-0001